ALSO BY JANET BAILEY

Keeping Food Fresh
Chicago Houses

THE GOOD SERVANT

Making Peace with the Bomb
at Los Alamos

JANET BAILEY

SIMON & SCHUSTER
NEW YORK   LONDON   TORONTO
SYDNEY   TOKYO   SINGAPORE

SIMON & SCHUSTER
Rockefeller Center
1230 Avenue of the Americas
New York, NY 10020

Copyright © 1995 by Janet Bailey
All rights reserved,
including the right of reproduction
in whole or in part in any form.
SIMON & SCHUSTER and colophon are
registered trademarks of Simon & Schuster Inc.
Designed by Edith Fowler
Manufactured in the United States of America

10 9 8 7 6 5 4 3 2 1

Library of Congress Cataloging-in-Publication Data

Bailey, Janet.
   The good servant : making peace with the bomb at Los Alamos / Janet Bailey.
      p.   cm.
    1. Atomic bomb—New Mexico—Los Alamos—History.
2. Nuclear weapons—United States—History.   3. Deterrence (Strategy)—United States—History. 4. Peace. 5. Cold war.
I. Title.
QC773.3.U5B35   1995
507'.20789'58—dc20   95-22737
                                         CIP

ISBN 0-684-80939-7

For Sue Davis

CONTENTS

	Prologue: A Different Country	13
1	The Plateau	21
2	The Pacific	45
3	Outside Las Vegas	64
4	Baksan, Russia	93
5	VA Hospital, Albuquerque	119
6	Mars	139
7	Operation Morning Light	165

Technology is a good servant, but a bad master.
　　　　　—Freeman Dyson

PROLOGUE

A DIFFERENT COUNTRY

> For with much wisdom comes much sorrow;
> the more knowledge, the more grief.
> —Ecclesiastes 1:18

ON JANUARY 13, 1993, I REALIZED THAT THE COLD WAR was over. I was in Amarillo, staying in a motel so dismal and malodorous that I never let my bare feet touch the green shag carpet. I woke up before dawn and drove eastward in the winter darkness. Just ahead, the thousand sodium lights of the only nuclear weapons assembly plant in the United States hovered like a swarm of yellow fireflies above the broad, flat wheat fields of the Texas panhandle. I joined the miles of cars, nose to tail, that flowed down a two-lane farm road toward Pantex. For thermonuclear warheads, this was Motor City, the factory where all the parts and all the history of weapons of mass destruction were bolted together into real hardware. I showed my ID to the guard at the gate, and he waved me through into the bomb factory.

PROLOGUE

Pantex had always been a sanctum sanctorum, absolutely out-of-bounds to anyone without a security clearance. But that January, there was a new task and a new policy to go with it. Instead of putting all those bombs together, the line workers were now taking them apart one by one, piece by piece. Inside the stout concrete walls, behind gargantuan blast doors, the weapons lay on wheeled gurneys. The technicians wore lab coats. They probed the bombs with x-ray machines and gleaming instruments in a rich parody of surgical procedure in which the patient was put to death.

Pantex was now a *dis*assembly plant, and so, the administration parted the curtains of secrecy and organized an open house.

Inside the visitors center that January day, a Pantex public relations officer fluttered around the press corps with the bright nervousness of the mother of the bride. Was the coffee hot enough? Did everyone have a dosimeter badge and color pictures of the B-61 fifty-kiloton gravity bomb?

Television crews crowded up at the front of the conference room and switched on their cameras with gum-chewing, been-everywhere, seen-everything boredom. Reporters shot questions at the lineup of Pantex employees gathered to greet them. Most of the questions nagged away at the troubling quantity of plutonium piling up on Amarillo's back porch.

Wait a minute. It was at that moment I saw that all those nuclear warheads, the nightmare megatonnage stockpiled and deployed, East and West, no longer represented doom and pestilence. Nuclear firepower, the stock-in-trade of Cold War politics, was not going to kill or maim every living thing on the planet, probably not this century anyway. The deadly nuclear arsenal, source of a million nightmares, had been transformed almost overnight into a solid waste management problem.

Pantex dismantled the oldest ones first, weapons built in the late fifties with nasty stuff like lead and cadmium. Through a maze of health and safety regulations, workers unmade the bombs with the hushed precision of watchmakers. Then they had to find ways to deal with the hazardous materials and all the other rubbish produced in the process.

It was necessary to disguise the secret shapes of nonnuclear weapons components, "demilitarize" them, before the parts were disposed of. Some nonnuclear parts were smashed to bits with a 300-pound forge hammer. Kevlar/nylon parachutes had to be shredded, but they were packed to the density of oak so they could squeeze inside a bomb casing, and they quickly blunted an ordinary metal blade. Pantex used water jets mixed with abrasives to slice through the chutes, "like bologna," an engineer said. Chemical explosives extracted from the warheads were burned to ash in an open field.

Once or twice a month, Pantex delivered carloads of aluminum scrap to a local recycler. At Tri-State Metals' main yard in Amarillo, car batteries, hubcaps, lawn chairs, household appliance parts, and aluminum cans rose up in ragged hills of junk. Every few minutes, battered pickups rolled in off the street with full cargoes of jetsam.

Tri-State Metals had an outpost, a big shed a few miles away where Pantex sent their bomb parts. All the fragments of aluminum went into a smelter where they were burned into liquid fire and poured into crude molds, "remelt sows." The remelt sows were sold to secondary mills. At the secondary mill, the aluminum was purified in giant furnaces, cast into ingots, and sold to manufacturers.

PROLOGUE

Most of the aluminum from dismantled nuclear weapons—no, it's not radioactive—was turned into auto engine blocks. Some of it ended up in homier circumstances, as barbecue grills and lawn mowers. Plowshares for the suburbs.

TWO YEARS BEFORE MY VISIT TO THE TEXAS PANHANDLE, I had begun the writing of a book about Los Alamos National Laboratory, the firstborn of all the institutions that built the world's nuclear arsenals. The laboratory was an engine of the Cold War, and the story of Los Alamos reproduced in miniature the eccentric evolution of the conflict.

The Cold War was still hot when I started the book and I wanted to know who was making nuclear weapons, what that job was, and why the weaponeers chose to do it. I wanted to know how the parts added up to the whole, to the omnipresent threat of annihilation. I had met a lot of weapons scientists and talked a lot of deterrence theory before that frigid day in Amarillo. I had followed the work of Los Alamos scientists whose projects, while they were certainly the fruits of the weapons program, had fallen far, far from the tree. I had even walked the desert floor where a thermonuclear device was buried, waiting for detonation in what would turn out to be one of the last bomb tests of the century.

Then the meaning of the Los Alamos story changed.

There, the as everywhere, the end of the Cold War came like the climax of a play, illuminating all the preceding action. The fifty-year rush of events suddenly ceased, permitting the audience to consider the meaning of what had happened. The characters themselves were left triumphant or confused or bleeding on the stage, but, to the observer, the story became self-contained.

Of course, we had all been characters in the Cold War.

PROLOGUE

We paid for the war with our money. We profited from advancements in technology, accelerated by Cold War competition. Most of us agreed to the primitive parity of nuclear deterrence, the suicide standoff. We had all felt the paralyzing touch of nuclear fear. We had all been at risk—hostages. We were all survivors.

As the Cold War drama closed, there was time to appraise the strategy of nuclear deterrence and its ferocious offspring, the arms race outside the atmosphere of terror. There was an opportunity to reexamine and perhaps to reinvent the institutions that had fueled deterrence.

We still owned the vast apparatus built to fight the war. We owned the Stealth bombers and the Space Program, the supercomputers and the nuclear reactors. We owned a lot of real property, like Los Alamos National Laboratory, and we owned the conundrum of how to transform the structures of war into the servants of peace.

The first impulse might be to rid ourselves straightaway of all the trappings of megadeath and shut the nuclear complex down, but there were those with a different vision of the future. The nuclear Cold Warriors, with one in the Win column, argue that nuclear weapons science is a fundamental survival skill, and that winning the war is no reason to stop investing.

So, in the midst of sorting through Los Alamos's nuclear machinery, its veiled and stigmatized past, I found another mission, which was to assess its alternate futures.

AS SOON AS TWENTIETH-CENTURY SCIENCE BEGAN TO SEE the boundless destructive power that lay within the atom, physics was suddenly a matter of pressing interest to heads of state. In 1943, the United States Army built a secret laboratory at Los Alamos.

PROLOGUE

Made of wood and tar paper, mud and barbed wire, the laboratory encompassed nearly all there was of nuclear physics at the time. It was remote, almost unapproachable, atop a New Mexico mesa, protected by guns and silence.

The enterprise at Los Alamos drew together a constellation of great minds: Fermi, Bohr, Feynman, von Neumann, Segrè, Rabi, Teller, Bethe, Frisch. The lab, or Project Y, as it was called, was an experiment—an urgent, heady, life-or-death rush into applied science, founded in the fear that Nazi Germany might produce its own atomic bomb and hold the world hostage. Nuclear physicists had been aesthetes, and now they were soldiers.

Construction of two nuclear reactors, one in Oak Ridge, Tennessee, and one in Hanford, Washington, began almost simultaneously. These early installations processed enriched uranium and plutonium, elements that could be coaxed into runaway chain reactions. The lab and the supporting nuclear plants were part of the Manhattan Engineering District of the War Department, the Manhattan Project, for short.

Within twenty-eight months of its founding, the scientists at Los Alamos knew how to make a weapon—a usable but crude, five-ton lump of lead, steel, and inspired improvisation. It was a fission bomb, fueled by the energy that burst out of an atom's nucleus when it fractured. By August of 1945, there were two atomic bombs in the world and both of them were used on Japan.

The same year, J. Robert Oppenheimer, director of the Los Alamos lab, gave a speech in which he said that the bomb "has led us those last few steps to the mountain pass; and beyond there is a different country."

Nuclear weapons did turn the world into a different

country, one in which war might extinguish history and not simply change it.

Los Alamos built more bombs, and the archipelago of factories that manufactured nuclear fuel and machined the bomb's structural anatomy more than doubled in size in only a few years. The Pantex plant in Amarillo, originally a conventional bomb plant, was added to the nuclear complex after it was decided that the United States was going to build more than a few of these weapons.

Long after World War II was over, Los Alamos remained a conclave—hermetic, anachronistic—still on a war footing, only now the enemy was the Soviet Union. The lab and the surrounding community were still isolated, literally locked behind barriers on a federal reservation.

In the fall of 1949, the Soviets exploded their own atomic bomb, ending the American monopoly. Scientists at Los Alamos responded with an even more arcane and difficult trick of nuclear physics: they built a thermonuclear weapon, a hydrogen bomb. Where the atomic bomb was based on splitting a heavy nucleus—*fission*—the power of a hydrogen bomb came from forcing two light nuclei to merge into one: *fusion*. The Soviet Union answered with its own thermonuclear device a year later.

The Cold War punch and counterpunch of nuclear deterrence continued for four decades, and Los Alamos prospered in pursuit of an open-ended mission to push nuclear weapons technology as far as it would go. This lab never worried about a budget. The program had a blank check and used it.

From the beginning, weapons physics at Los Alamos relied on the skills of all kinds of scientists. When the weapons designers needed metals with the properties of glass, Los Alamos imported chemists and metallurgists to create them. When the bomb makers needed to gauge the dynamics of neutrons, the Los Alamos mathematicians invented calculations to do it. As the laboratory matured, the support sciences enlarged and elaborated and some even left bomb making behind. By the nineties, 40 percent of the lab's budget went for what one former bomb maker called "the lint off the weapons program," research projects ranging from microbiology to space exploration, particle physics, AIDS research, cosmology, genetics, mathematics, and dozens more. The secret city had become an open campus, but still each project could trace its genealogy back to the grim business of weapons research that still remained the lab's heart and soul.

By 1992, Los Alamos, its sister weapons laboratories, Sandia and Lawrence Livermore, and the string of weapons plants across the country had crafted a stockpile that overwhelmed the imagination with its omnipresence, its deadly efficiency.

And then the Cold War was over. The two great powers holstered the guns of mutual extinction. It was a revolution of the possible, a redrafting of the future. Armageddon was an artifact of the past.

We were suddenly in a *different* different country.

CHAPTER

1

THE PLATEAU

> All professions are conspiracies against the laity.
> —George Bernard Shaw, *The Doctor's Dilemma*

JAMIE GARDNER GOT OUT OF HIS PICKUP WHERE ROUTE 4 finishes its first corkscrew rise above the Pajarito Plateau. He had parked at the edge of a cliff on a sandy half-moon of land, a lookout point for rangers to scan the surrounding forest for signs of fire.

Jamie wanted to show me the view. We were on the eastern slope of the Jemez range, a ring of mountains hollowed out by a volcano a million years before. The mesas below us made up the plateau, positioned like a stair step between the floor of the Rio Grande valley and the peaks above us. Atop the mesas, on five fingers of land separated by deep, sheer-walled canyons, was Los Alamos—the laboratory and its sidekick town: birthplace of the atomic bomb and still a wellspring of nuclear weapons technology.

In northern New Mexico, mountains jutted out of a desert of red rock and ground-hugging piñon trees. Between villages, the roads were long and silent. Los Alamos National Laboratory was by far the largest enterprise in the region, but it operated in cultural isolation. It still kept secrets and these secrets fostered tales like the one about a farmer in the village of Chimayo who bred mammoth burros and grew chile peppers. He also raised homing pigeons and they always lost their way in the skies above Los Alamos. Buried in these stories you could always find some version or another of the Invisible Ray Theory. Whatever it was up to, Los Alamos commanded attention.

It was the autumn of 1990 and the Nuclear Age was forty-five years old when I first came to Los Alamos. My generation and I had lived our entire lives with the Bomb, a literal gun to our temples. Los Alamos was the very source of the Nuclear Age, a place where I could look down the barrel and see who was slipping bullets into the chamber.

I arrived with routine English-major, liberal-politics equipment: I was uneasy with militarism and had a tin ear for mathematics. I was no firebrand, but I had put in my time, working the neighborhoods for losing presidential candidates and raising my voice against the war in Vietnam. A Chicago cop had rapped me over the head with a billy club during the 1968 Democratic convention and given me a special outlook on the abuse of power.

I thought of the Bomb as something wicked and out of control. I thought that every weapon added to the stockpile was another black ace stacked on a house of cards, a structure that, as it grew in size, was more and more likely to collapse and bury us all in a final, poisoned grave.

THE PLATEAU

I truly wanted to know if there was a reasonable explanation for this.

So there I was, standing on a ledge above Los Alamos with Jamie Gardner, the first lab scientist I found to talk to. Jamie looked like the leader of a mild-mannered motorcycle gang. He was stocky and rough cut in his worn flannel shirt and hiking boots. On the belt slung low around his hips, he packed a rock pick and a magnifying glass.

Jamie was a geologist. Geology was one of the tributaries that fed the laboratory's racing stream of weapons science. In the sixties, when they started drilling holes in the desert for underground nuclear tests and measuring the artificial earthquakes that followed an explosion, weapons designers had needed geologists. Like the lab's chemists and mathematicians, once the geologists were on the laboratory staff, they stayed there, pursuing projects of their own while waiting for the bomb makers' next call. Jamie's current project was an attempt to tap the Jemez volcano's buried reservoir of heat to use as clean energy.

The earth under our feet had a history of its own and Jamie could see it all around us.

"The volcanic eruptions that blew the top off the Jemez must have looked like hydrogen bombs going off, sending columns of hot ash miles into the air, mushroom clouds," Jamie said. "Molten ash spilled out and flowed downward. It was viscous, like a moving wall of pudding. It crept into the valley and cooled into these mesas."

Jamie's only brushes with weapons physics were trips to Nevada to evaluate burial sites for radioactive waste. When he talked about weapons at the lab, he talked about what was "inside the fence." He was as much an outsider as I was.

Jamie told me that he had spent months, years,

walking the Jemez, making geologic maps of the area from what he called "ground truth," the accumulation of facts he could see, smell, and touch by simply being there, inside the landscape.

"I look for faults and folds and abrupt changes in the age and composition of the rocks. I record locations and take samples of soil to make an accurate map, but I also formulate a mental model from the evidence at my feet. I reconstruct the past: What series of events happened that would have ended up looking like this?"

Jamie thus presented me with a metaphor for the job I was about to undertake. I was going to look for the laboratory's ground truth. I would dig around in the pure science, political theory, and technical sorcery that were the roots of nuclear deterrence.

That day, Jamie and I got back in the truck and drove higher into the Jemez. Fifteen minutes up the road, we were deep in Rocky Mountain montane forest. In this country, where a few hundred feet of altitude simulate a few hundred miles of latitude, the cactus and juniper quickly gave way to towering ponderosa and dense stands of pencil-thin aspen. And fall came earlier at these elevations. It was bright and hot below us, but here, beneath the dark evergreens, night frosts had already colored the scrub oak chrome yellow.

DURING ONE OF MY EARLIEST TRIPS TO LOS ALAMOS, I was following a road marked with blue signs saying "Explosive Trucks This Way" when I made a couple of left turns and found my car aimed at a guard station flanked by a ten-foot-high chain-link fence with scrolls of razor wire at the top. I thought, *Uh-oh*.

The guard ducked down to smile through the car window and waved me through into the fenced area.

"Just hang a U right here. It'll get you back on the main road."

No guns drawn, no body searches, no sweat. Here was what Jamie called "the fence." The razor wire was unusual—this was the first time I had seen it. Los Alamos did not rely on hyperbole for its security. For the most part, a kind of visual silence protected the lab's forty-three square miles: Los Alamos National Laboratory simply did not address itself to the outside world. Even the signs were uninformative: "Tech Area 55," or "SST-10." The lab was one of two places in the United States where nuclear weapons were designed, but it looked like an underfunded community college or a factory that couldn't make a profit. Prefabs a little larger than portable classrooms clustered around the doors of truncated office blocks. Warehouses and smokestacks could be glimpsed behind a screen of trees: they could have been fabricating tin cans instead of bomb parts. Most of the buildings seemed ad hoc, like somebody got an idea, threw four walls around it as fast as possible and got to work.

Only the plutonium facility told a story: it looked like the state penitentiary.

You needed a security clearance to get inside most of the Los Alamos laboratories, but anybody could attend the every-Thursday salvage sale when discards from laboratories and workshops were available to the public.

This high-tech yard sale posed many mysteries and encouraged speculation. Of course there were many ordinary items: towering machine shop tools, drills, circular saws, barrel-shaped motors; file cabinets and computer parts, monitors and motherboards; great lengths of pipe and flexible ductwork. More compelling were the electronic consoles heaped on the asphalt, their knobs and dials and digital readouts not streamlined but boxy, utilitarian.

The day I went to the sale, there was a hooded Dumpster the size of a tollbooth with "Non-Radioactive Trash Only" stenciled on its side. There was a stack of wire animal cages, three bucks apiece. Among hillocks of steel scrap, Plexiglas slabs, coiled cables, and gutted aluminum casings, I found a large white erasable message board still covered with grease-pencil writing. Individually, the words were understandable; together, not: "Natural Fracture Reservoir"; "2 for Process Field Operation Strategy."

I saw an inexplicable five-foot-tall beige metal box whose only features were an empty niche and a rheostat marked "DOSAGE."

FOR THE FIRST SEVERAL MONTHS IN THAT FALL OF 1990, I poked around in the sciences outside the fence where there were no secrets. The weaponeers were not answering my phone calls. Finally, I met a friend of a friend and he had a friend and I arranged to meet Delmar Bergen in the cafeteria on a cold winter day.

The cafeteria was a rare instance of grace at the lab. It had green plants and skylights and a wall of windows with a view of the Sangre de Cristo Mountains.

The people in the cafeteria had a look all their own. The scientists were definitely uncool—the short-sleeved polyester shirt was standard. Many of the staff appeared to have fallen out of bed at seven forty-five and dressed in time to make an eight o'clock class.

A number of years before, when airplane hijacking was a concern, the laboratory issued a memorandum to the scientists, giving them pointers on how to conceal their identity from terrorists. One of the suggestions was that they make sure they were wearing matching socks.

Bergen was more stylish than the average. He was a small, springy man in his sixties. Except for the network of lines that scored his face and his steel gray hair, he looked twenty years younger: he bounced around in jeans and Reeboks, moving like a gymnast.

"I was born about the time the neutron was discovered," Del said. The neutron would turn out to be the subatomic bullet that smashed the atom and forced immense, hidden energies out of the nucleus.

The physicists who built the first atomic bomb were fighting World War II. When Del joined the lab in 1957, the lab's mission was to arm America for the Cold War.

The Cold War was constant and dangerous. Only months before Bergen signed on at Los Alamos, Russian troops had crushed an anticommunist revolt in Hungary, doing battle with civilians in the streets of Budapest. In the first years of Bergen's life at Los Alamos, the Russians launched Sputnik, conducted the world's first intercontinental ballistic missile test, built the Berlin Wall, and exploded a fifty-megaton hydrogen bomb, the biggest in history. A U.S. spy plane was shot down over Russian territory, and guerrillas, backed by the CIA, invaded Cuba at the Bay of Pigs.

"In those days, what we did was automatic," Del said. "During the weapons buildup between fifty-seven and sixty-two, the mentality was 'If those Russian crazies come after us, we'll blow them to oblivion.' The U.S. built all the firepower it possibly could: you built it and you put a nuke on it—every kind of weapon we could conceive of. It was assumed to be in the interest of the United States—no public relations necessary."

Del Bergen grew up on a western Kansas wheat farm, where he was tooling around on a tractor by the age of twelve. His father was a math teacher. His mother had studied to be a lawyer, although she never practiced.

"When I was in high school, I wanted a pickup truck

and I wanted to play center field. I went to college to be on the baseball team. I just goofed off for the first two years."

In 1947, when Del was still in high school, there were thirteen atomic weapons in the world, all built at Los Alamos. Bomb parts were cast by hand. It took a specially trained thirty-nine-man team more than two days to fit the parts together. The technicians there were the only people in the world who knew how to handle plutonium and uranium 235—both are pyrophoric: when they are machined, the dust catches fire. Los Alamos theoreticians were the only people in the world who knew how to calculate the point at which a mass of uranium 235 is exactly the right size and shape to start a runaway chain reaction.

Los Alamos was the major leagues of physics. Atomic bombs were a symbol of national strength.

"When I was in college, I wanted to drive out to the desert and see one of them go off," Del said.

Before his education was over, Del had recovered the earnest dispositions of his upbringing. He married his high school sweetheart and returned wholeheartedly to the Christianity in which his parents had raised him. He discovered an enthusiasm for physics. Del and his wife were already raising two children when he signed on at Los Alamos. He had been working his way through graduate school, and now the lab was going to pay him to finish his Ph.D. The family moved into assigned housing, a three-bedroom in the suburban-barracks style that still survives near the center of town.

It was an extraordinary town. When J. Robert Oppenheimer, the lab's first director, picked the site it was because it was miles from nowhere, an easy place to hide. During World War II, the very existence of Los Alamos was secret. Harold Agnew, a Project Y scientist who would become the lab director in the seventies, recalled that during the war his parents were embarrassed, because while everyone else's kid was in uniform, they didn't even know where he was, let alone what he was doing. Until the mid-fifties, the whole place continued to be owned and operated by the federal government. Every inch of it was locked up behind miles of fencing. You needed a pass to get through the guard gate, past the armored tanks that flanked the road, and into town. You needed clearance to work at the drugstore or the gas station. Everywhere but Los Alamos was the "outside world." It was small and safe and the Zia Company that serviced the lab would also come and fix the plumbing in your house. The gate was thrown open in 1957 just a few months before Del Bergen arrived.

"I missed out on the era when you could exclude unwanted relatives. They used to call the guard station the 'mother-in-law gate,' " Del said. "I was second generation after the Manhattan Project. In weapons physics, we felt we were an elite, that what we were doing was important."

In 1957 there were six thousand nuclear weapons in the American arsenal. The laboratory was no longer the singular center it had been. A dozen production plants in as many states churned out nuclear fuel and mass-produced components. A rival design lab, Lawrence Livermore, in central California, was in operation.

The broad strokes of bomb physics were in place. What remained was to coax the basic physics into clever, efficient packages that could be relied upon to explode at the right time and on target.

"I first joined a combination engineering/physics group working with heavy metals," Del said. "Even though right after the war the work of the weapons complex had been parceled out to different factories all over the country, Los

Alamos still had the capability of doing all the parts. We often went to full assembly on test devices. In fact, the only time we didn't was when we were uncertain about the safety of the design. Then we would wait until everything got to the proving grounds in Nevada before we put the high explosives on. Back then, Los Alamos was still a garage operation. It was all hands-on. You could get an idea and just cut and shoot." In other words, he could walk a test all the way through: carve the heavy metal for an implosion device and then blow it up to see how it worked.

This kind of thing suited Del nicely. He was a tinkerer, not a desk man.

When Del talked about nuclear politics, he sat up straight and chose each word deliberately and carefully. He had spent a lifetime defining and measuring his place in the science of megadeath. The logic of deterrence demanded thoroughness and precision.

When he talked about physics, he leaned back and let it roll. "Bomb design is taking ten kilograms of chemical high explosive in the primary explosion to get fission energy worth ten million kilograms. Then you take a fraction of the ten million to enhance production of more energy in the secondary. It's obvious we know how to build that up to the equivalent of ten billion pounds of chemical explosives. The primary starts in the realm of energy of a campfire or a combustion engine and the secondary achieves the temperature of a star. All this happens in a volume of space no bigger than your living-room couch."

To design a nuclear weapon is to orchestrate pandemonium, to tame the behavior of 10^{24} atoms (a literally incomprehensible number), atoms that gyrate through turbulent fields of force, that merge or burst, kicking out fragments and violent radiations.

Textbook physics alone cannot tell you how to build a bomb. "First principles would say, I've put in enough energy, the reaction should occur," said Del Bergen. "It doesn't happen that way." After thousands of experiments and hundreds of nuclear tests, the scientists know what goes bang and what does not.

In the early years, Del was sent out for a month at a time to hotdog around the lab's test site in the South Pacific. The Marshall Islands were two parallel chains of smooth coral caps strung out along the equator northeast of New Guinea. The United States tested dozens of fission and fusion devices there between 1946 and 1962.

"Los Alamos worked at Eniwetok atoll. We saw a lot of leftovers from the Japanese occupation, like shells and sunken concrete boats. Fred and Elmer Islands were for housing and for landing planes. The days were hectic but in the evenings we played cards and talked. Our atoll surrounded a big saltwater lake protected from ocean tides. The shallow water was like an observatory for sea life. I would snorkel and see sharks and manta rays. I remember watching tuna herding small fish that got washed over the coral head and into the open water when the waves were high.

"The shots themselves were always exciting. I don't know how many you could see before it would became routine. I remember one mushroom cloud, miles away, that went up and up to maybe fifty or eighty thousand feet and then it started spreading, fast. Pretty soon our heads were thrown all the way back and we were looking straight up at it. I thought, 'That's a *lot* of radiation up there.'"

During his career at the lab, Del had moved through the ranks, starting with engineering, graduating into thermonuclear weapon design, and then into management.

Weapon science tended to wear people down. The physics was grueling, and the punishing, full-tilt timetables of nuclear tests rarely let up. Few scientists spent their whole careers in the design division.

Los Alamos weaponeers often found themselves at the confluence of science and foreign policy. Del served tours of duty first in Washington as a science adviser in the Department of Defense and later in Geneva, participating in international arms control negotiations.

Like most members of the weapons community, Del believed nuclear weapons had done their job of deterring a global war, a hot war, between East and West. "The development of all the tactics and all the tools looks like preparation for war, but those fundamentals are the strategy for preventing war," he said.

ONE OF BERGEN'S ANTECEDENTS, THE DANISH SCIENTIST Niels Bohr, was a sweet-tempered genius. He endured the Nazi occupation of Denmark until 1943, when it became clear that his status as the greatest living theoretical physicist could no longer protect him from arrest. When he came to Los Alamos to participate in the Manhattan Project, he bore witness, in this protected community, to the savagery of the conflict in Europe. Bohr put this question to Oppenheimer: "Is it really big enough?"

Was the atomic bomb terrifying enough? Would the cost of global war now be so high that no great power could risk the first shot?

Bohr grasped that such a bomb would change the art of war. Instead of a contest of strength, it would be a bloodless ledger of accounts, balancing terror against political profit.

What was clear to Bohr was not immediately obvious

to military strategists. When America's leaders finally possessed the ultimate weapon, they treated it like TNT, a bigger, better stick of dynamite. It was as if this infernal machine had slipped back from the dark plain of the future into their hands and they had only the antiquated patterns of the past for blueprints.

The bombing of Japan was an extension of the intensive air raids that had destroyed cities in Europe and Asia throughout World War II. In the postwar forties, the Strategic Air Command (SAC) was the custodian of nuclear policy. Bombing civilian populations was SAC's sole nuclear battle plan.

Among SAC's armory was the Mark 17, a thermonuclear gravity bomb with a yield in the range of fifteen to twenty megatons. Los Alamos lab put the Mark 17 together. It was the most powerful nuclear weapon the United States ever built. The bomb was five feet in diameter and weighed 42,000 pounds.

Del Bergen was showing me a defanged version of the bomb that was on display in the National Atomic Museum in Albuquerque. Del said, "It was a time of confrontation and belligerence and we wanted the biggest bang. We believed massive destructiveness was the way to scare the Soviets off. We were still operating with the psychology of World War II: if *some* of us survive, that will be enough."

It was an ugly son of a gun: fat, blunt, ashy black. The Air Force needed two cranes to load it onto a B-36 bomber. If it was possible to quibble about whether an atomic bomb was deadly enough for deterrence, this hydrogen bomb, burning with the heat of nuclear fusion, was truly a doomsday device.

The Mark 17 only made sense as a city killer, the kind of all-out, no-mercy, hellfire and fallout ordnance whose survivors would envy the dead. All of those twenty megatons were meant to frighten the Soviet Union away from meddling outside its already expanded geography of domination. The policy called "massive retaliation" allowed for the use of atomic bombs against the U.S.S.R. either to answer or to preempt communist incursions into Western Europe or Asia. The United States would no longer send ten thousand boys to the front with rifles. Instead, it would send the Mark 17.

If it was going to influence Soviet conduct, a threat had to be authentic. In the case of the Mark 17, the logistics were a problem. A B-36 aircraft was not a precision instrument. Like all bombers, it was vulnerable to air defenses and it operated poorly in bad weather. Of course, at least part of the point of twenty megatons was that, if it hit two miles off, the bomb could still cripple Moscow. But, with Mark 17s on board, the range of the B-36 was cut from nine thousand down to two or three thousand miles, lumbering along at no more than a couple of hundred miles per hour. In order to drop a Mark 17 on target, the bomber had to be parked on Russia's doorstep. If a B-36 was close to the Soviet Union, the Soviet Union in turn was close to the air base. The B-36 was a sitting duck.

More important than its tactical shortcomings were the B-36/Mark 17's logical faults: Would the Russians really believe that a president was going to react to a skirmish in some German province by incinerating the cab drivers, bakers, professors, nurses, schoolchildren, and poets who filled the streets of Moscow? The peculiar political delicacy of the situation could be seen in SAC's designating as ground zero not the Kremlin but an electric power station one mile away—even the Strategic Air Command was squeamish about killing Russian mothers and children.

Another tale of the Mark 17: One spring morning in 1957, a B-36 was flying a Mark 17 bomb from Biggs Air Force Base in Texas to Kirtland in Albuquerque. Some press accounts say that a crewman unwittingly leaned on a release mechanism. However it happened, as the plane approached Albuquerque forty thousand pounds' worth of bomb fell through the bomb-bay doors. The Mark 17's chemical explosives blasted a hole in New Mexico's desert floor twenty-five feet wide and twelve feet deep, leaving faint traces of radioactivity inside the crater.

A bomb was the promise of apocalypse, but it was also a widget. It had to be handled—lifted, trucked, stacked, and therefore possibly dropped on the floor. It had to be built to go off when and where it was told and definitely not to go off on any other occasion.

The Mark 17 that fell did not do more terrible damage because its capsule of nuclear explosives was out of harm's way on board the B-36. It was a design fundamental: don't carry the match next to the tinder.

"A weapons designer had to consider all possible circumstances," Del said. "The first consideration was a bomb's intended use; then we had to think about accidents, and we had to think about misuse."

The United States put nuclear weapons in Europe early in the fifties. As soon as they were on foreign soil, security was a problem. Designers at Los Alamos had to introduce mechanical controls so that, even if "unauthorized personnel" managed to swipe a nuclear weapon, they couldn't detonate it.

Weapons designers never ran out of work. All the nuclear arms on display in the museum were just a small sample of the long chronology of weapons systems—more than a hundred—developed during the Cold War. And for every weapon on display, there were a multitude of false starts, aborted tests, failed prototypes.

The year Del Bergen had joined the staff at the laboratory was the year the Air Force nearly bombed Albuquerque. It was the first year of rampant growth in the nuclear stockpile, the start of a decade that would see more weapons and more megatons than before or since.

"It did not happen the way you would like to think, with strategists rationally picking out the components of an arsenal to fit a coherent theory of deterrence," Bergen said. "The technology sprouted up and the strategist came along behind trying to make sense of it. The technology to hit military targets preceded the strategy to do war-fighting rather than city-burning."

The Mark 17 was succeeded by compact, efficient bombs that were mobile, accurate, and relatively "clean," that is, less radioactive. The modernized weapons allowed for a nuclear doctrine subtler than SAC's state terrorism. The new policy, called "flexible response," armed all the branches of the military with weapons suited to combat on the ground.

Bergen was on the team that designed the exemplary B-61. The B-61 bomb was a jewel of streamlined clockwork engineering, a slim, sharp-nosed bullet about eleven feet long, with four tail fins arranged like feathers on an arrow. Inside a lustrous aluminum body, the bomb had sense organs—radar, trajectory timers, inertial switches—that fed back information to the arming mechanisms: the B-61 had to be aiming in the right direction, tracing a predetermined path, and falling at the expected speed, or it shut itself down. It was equipped with a lock that prevented the weapon from being armed until the correct code was entered. A B-61 could be customized to detonate in the air or at surface level. It was built to withstand shock, so that it could survive hitting the ground and then go off after a time delay. A far cry from the ungainly Mark 17, it offered the military a host of options.

A B-61 could neutralize a massed battalion of tanks. It could blast away a Russian air base instead of razing the city of Leningrad. The B-61 was a realistic threat, and therefore, the theory went, the B-61 was a more effective deterrent.

"As a political weapon, nukes are only as good as the enemy believes them to be," Del said. "You have to be prepared to fight a war to deter it."

Del Bergen retired in 1991, the same month the communists closed up shop in Moscow. He escaped the discontinuity and personal upheaval about to befall his colleagues in weapons work.

JAMES (NICKNAMED JAS) MERCER-SMITH AND I WERE ON the upper level of the Oppenheimer Study Center, lounging on couches arranged outside the meeting rooms. One floor below us was the Los Alamos laboratory library, a large collection, open to the public. More than open, it was as cheerful and democratic as any public library in any small town. The reference staff would scurry around, tracking down obscure books and pamphlets for me, paying no mind to my nonstatus at the lab.

Mercer-Smith was sober and decorous. In his attire, he struck a balance between student and professor, wearing suit jackets over tieless, unbuttoned dress shirts. He had a pale, oval, boy-angel face—this was not the face of a kid who ever smoked cigarettes behind the schoolhouse. Astrophysics was his area and his approach to it was a distinctly cerebral one. "I liked building mathematical models of how stars worked. I was not all that interested in looking at them through a telescope. You have to look at them at night when it gets cold and dark." Jas was in his late thirties. He earned his doctorate in astrophysics at Yale, then went to Harvard as a postdoc to inquire into the formation and thermonuclear ignition of low-mass stars. "I looked at how stars burned nuclear fuel. Now I get to do the burning myself. Of course, these stars are much, much smaller."

In the basement of the Study Center where Jas and I sat in conversation was the library of secrets, archive of all the classified texts.

The record of the Bomb's technical genesis was apparently patchy. The physicists who designed the earliest weapons did not stop to narrate what they were doing. I was told, for instance, that when scientists wanted to study long-term health effects from the Hiroshima and Nagasaki bombings, "we could not backtrack with precision because we had lost information on exactly how the first bombs were made."

Jas once searched out the original paper by Edward Teller and Stanley Ulam that solved the puzzle of how to make a thermonuclear bomb. "It is in the best interest of any temporary custodian of a secret document to destroy extra copies rather than risk security problems. I found no more than five copies of the Teller-Ulam document stuck in various safes around the lab."

In the absence of a codex, skills and secrets were passed from person to person, on the job. Rookie designers learned by doing. "It's illegal to teach this stuff anywhere else," Jas said. "Two years of apprenticeship makes you useful. After five years you can do work without hurting yourself. After four or five test shots, a designer knows how to do the job. At fifty million a shot, that's a quarter of a billion dollars in training."

Mercer-Smith had worked in the Thermonuclear Applications Group, X-2, since 1983. Jas specialized in what he called "high-risk" ideas—with no apparent immediate practical application. "I don't have any designs in the stockpile," he said. The weapons budget always had room for blue-sky inventions.

The work of designing nuclear weapons had changed a lot in fifty years. The first generation of designers had to put down their pens and pick up wrenches to get their bombs built. When Del Bergen's generation designed nuclear weapons, they could still get their hands dirty. Mercer-Smith's only tool was a supercomputer.

In 1943, when a bomb designer talked about a computer, he was referring to one of the many people, often Los Alamos housewives, who sat for endless hours doing arithmetic on mechanical adding machines.

Calculating the physics of a nuclear detonation pushed the tools of mathematics to their limits then, and it still did when Mercer-Smith's generation came along. Scientists had to measure the behavior of shock waves and the flux of infinitesimal particles through a knotted, seething amalgam of metal. From the time of the Manhattan Project, the demand for high-speed computing in weapons work dragged the computer industry in its wake. One of the first large-scale computers, the MANIAC (Mathematical Analyzer, Numerical Integrator, and Computer) was built at Los Alamos in 1952 to do the mathematics of thermonuclear weapon design. The weapons program always bought the biggest, fastest machines as soon as they were wired together.

"I work on a Cray that does a trillion multiplications

per second," Jas said. "I do more numerical calculations in an average month than were done in the history of the world before nineteen-seventy.

"We have thirty-five years of accumulated data from experiments translated into a half million lines of computer code that model the physics of nuclear detonations," he went on. He manipulated this data, using it to predict what will come out of a novel device—neutrons, gamma rays, and kinetic energy.

"It is like sitting down at a massive organ to play a Bach fugue. The trouble is, when you press one key, this key over here is going to sound different. It takes a lot of experience and intuition to make the music sound right."

The average book is equivalent to about one megabyte of coded information. For Jas, a single calculation could involve a gigabyte—enough information to fill a library. "I'll run hundreds of those on a difficult design. I have to look at the outcome graphically—you can't deal with the actual numbers. What you get is only an approximation. The codes lie. An experienced designer knows exactly how the codes go wrong, and can go on to make correct assumptions."

Even the most powerful computer cannot simulate the fast and fluid evolution of nuclear reactions. The best it can do is to define cells of time and space, and plot their movement through the rushing stream. The cells are like pixels in a newsprint photograph, dots that create only an illusion of continuity.

"Computer modeling is imperfect," Bergen had said. "But at least it slows the process down so you can look at it."

Weapons physics happen on an inhuman time scale. Jas said, "We count nuclear reactions in a time unit called a shake—there are one hundred shakes in a millionth of a second."

To get an idea of how finely sliced this unit was, picture the distance between New York and Los Angeles as one second. The first two inches correspond to one shake.

The exploding fission-fusion-fission chain of a thermonuclear explosion happens so fast that the bomb's aluminum skin is still intact when the reactions are finished.

ASKED HOW THINGS HAD CHANGED IN LOS ALAMOS, DEL Bergen remarked, "When I joined the laboratory, I was swept along by the current. When Jas came, he had to swim upstream." In 1949, opinion polls in the United States showed that nearly 60 percent of the public believed it was a good thing that the atomic bomb had been developed. In a 1982 survey, 66 percent thought it was bad.

Jas Smith grew up in the state of Washington, near Bremerton naval station. Both of his parents were naval officers in World War II. His mother was forced to leave the service when she got pregnant with Jas. His father left the Navy a few years later to become a teacher. For forty years, in the military and out, his mother was a nurse anesthetist.

Both Jas and his sister had learned the lesson of service from their parents—she grew up to be a nurse on the Navajo reservation. With Jas, Mr. and Mrs. Smith knew they had a special case: "When I was six years old," Jas said, "Dad played a game where he would flip cards as fast as he could, and I would keep a running sum—something I couldn't do today."

Jas attended private schools. He went to elementary school with Dominican nuns, and to boarding high school with Benedictine monks. As an undergraduate, he attended New College, in Sarasota, Florida. "They didn't issue grades at New College, which made the competition far more vicious. In four years, I visited the beach exactly two times in daylight. At Yale and Harvard, I would work from seven in the morning to midnight."

In the mid-seventies, he met his wife-to-be, Janet Mercer, in a summer program at Oak Ridge National Laboratory, another part of the nuclear weapons complex. She was working on a degree in chemistry.

Jas said, "I took the job at Los Alamos for a couple of reasons. For one thing, my wife could work here too. But I could also see that the best scientists in the United States were not going into the weapons program. In the Soviet Union, the best were *drafted* into weapons. Students of Zel'dovich, Russia's greatest physicist, published for a while, then vanished. The lack of symmetry was destabilizing. So here I am.

"A science like cosmology is aesthetically pleasing but it doesn't *matter* whether the universe is expanding or contracting. Weapons are real-world problems."

At home, Jas Mercer-Smith fixed most of the meals. "My wife is a chemist. She is a better cook than I, but she slowly measures every ingredient to the last microgram—it just takes too long. It's easier to do it myself."

Jas was in his kitchen one afternoon when he returned a call. "What we do is not civilized," he was saying. "We threaten to destroy people so that they will respond out of self-preservation."

I could hear the clank of dishes in the background. "I think about what I do every day," Mercer-Smith said. "I work on deterrence every day. If I am wrong about deterrence, the deaths of hundreds of millions will be on my hands."

Jas had come to Los Alamos because he passionately wanted to do what was right. The end of the Cold War meant abandoning nuclear weapons or redefining their use; it meant doing a different job than he had been doing much of his adult life.

What had happened in Los Alamos had been mythic. The Manhattan Project brought the demigods of science together to save the world by devising the means to destroy it, but nowhere in the town was there a monument to this signature event of our time. An attractive museum matter-of-factly related the story and its sequelae. A street sign designated the corner of Oppenheimer and Trinity, but there was nothing in the way of statuary. Neither was the end of the Cold War a moment of triumph or reward. There was no place to plant a flag.

Los Alamos's last Cold War nuclear test went off in Nevada in September 1992. No new tests were planned, no new weapons were contemplated. After a fervent fifty-year battle, the Cold Warriors had laid down their guns, and now they didn't know what to do with their hands.

By 1993, Jas, promoted to group leader of X-2, was mothballing most of the thermonuclear design work. "People in the division are starting to look for other positions," he said.

Jas had no plans to leave. His work had been nuclear deterrence, a brittle, dangerous, sometimes incoherent interim solution—it still is. "We don't even know exactly what we are deterring anymore," he said. "Who is the adversary? The threat of Armageddon may be gone, but chances of a single attack on the United States are greatly increased."

In weapons designers, I learned to recognize a characteristic cerebral, gloomy optimism, an acute awareness of their place in the world. Jas told me this story from Herodotus: The king condemns a merchant to death for some infraction. The condemned man negotiates with the king, saying, let me live one year and I will teach your horse to sing. The king agrees. When the merchant's friends point out that this is not much of a bargain, he replies that anything can happen in a year—the king could die, *he* could die, and it is just possible that a horse might learn to sing.

"Deterrence just buys you time," Jas said, "and you hope the world will come to its senses—that it will learn to sing—before nuclear weapons are ever used again."

CHAPTER

2

THE PACIFIC

> Extreme remedies are most appropriate for extreme diseases.
> —Hippocrates

FOR MANY PEOPLE, THE GEOGRAPHICAL CENTER OF THE Nuclear Age is Hiroshima, with its shadow corpses burned into the pavement by the awful heat of an American bomb.

The weapons designers at Los Alamos drew a different map. The boneyards of Europe and Asia defined longitude and latitude: the Somme, Stalingrad, Dresden, Auschwitz, the Gulag, Nanking, Bataan. Inside these territories, humankind needed no special tools to inflict suffering without limit.

Perhaps twenty million people—soldiers and civilians—died in World War I, and forty-five million more in World War II. In a roll call of these war dead, Hiroshima's and Nagasaki's took their place as the final quarter million.

And each death counted as a failure, a failure of civilization to safeguard its own.

After 1945, an ultimate showdown between the most powerful nations seemed so predictable that it already had a name: World War III. The technical capacity to kill, the sheer lethality of standing armies and their weaponry grew to the point of threatening every man, woman, and child on earth. Even so, the dead, slaughtered in rebellions, proxy wars, guerrilla fighting, massacres, and terrorist bombings—while still numbering in the millions—were only a fraction of the billions who could have died in the first hours of a nuclear conflict. For the true believers, there is only one reason we've been spared—because of the Bomb. This is the theology of nuclear deterrence.

"All weapons are terror weapons," Harry Berlijn said. Harry was a physicist in Los Alamos's thermonuclear design group, Jas Mercer-Smith's coworker. Harry was Eurasian, born in Indonesia. There were several foreign-born scientists in the design division, from Eastern Europe, from war-ravaged countries in Asia, some of them women.

Harry spent World War II in a Japanese concentration camp in Indonesia. "My father said not to fear the Japanese invasion. He said they would need us, the professionals, teachers, doctors and so on. 'Don't resist,' he said, 'there is always hope if you remain civilized. They will spare us,' he thought. We went straight to a camp."

Harry's father died there, under torture.

"Toward the end of the war, thirty people died every day—cruel deaths from starvation and dysentery," Harry said. "At the rate it was going, all of us were going to be dead in six months. We heard the atomic bomb had been dropped, and then there was peace. They had beaten and tortured us, then a week after the bomb, the Japanese were giving us rides in their jeeps. I know at least one guy whose life was saved by the atomic bomb."

Jas had made a point of introducing me to Harry, because he was part of a priesthood of memory within the design community, scientists whose own battleground biographies reminded everyone else of what was at stake. Harry was in his sixties. His looked like a sweet and pungent old age. Harry clarified for me a common trait I had observed among many weaponeers: they saw the world as a very dangerous place where the worst was always nearby, only waiting to happen. For them, the Bomb stood for a kind of tragic, last-ditch optimism, the hope that national leaders would choose self-preservation over anger. Unlike conventional war, these scientists are quick to remind us, nuclear war can kill politicians.

"There are other Hitlers waiting to be born. We are not civilized—we act from fear. *Assured* destruction, that's what's important.

"The Cold War was rough on weapons designers," Harry said. "We had to defend ourselves against moral outrage. But I didn't join the lab to make more Hiroshimas. I thought, 'We can stop war.' The Cold War was World War III. This is how I fought in the Cold War."

THERE IS ONLY ONE BAR IN LOS ALAMOS. LARRY MADSEN and I were sitting there, talking over a glass of wine at the end of the day. In the Bomb's hometown, nuclear war was no more real than anywhere else. Neither was it unreal.

"By 1992, they had cleaned out all the old food stores from the lab's bomb shelters," Larry Madsen said. "Mice had invaded everything, even the canned goods. When I first came to Los Alamos, employees and their families had a shelter designation. Later on, we didn't bother with that kind of civil defense because we knew the Russian missiles were just too accurate."

The Nuclear Age was almost fifty years old and no one on the face of the earth knew what would happen in a nuclear battle. No nuclear war had ever been fought. No one could predict the precise dimensions of such a war. Nuclear war was entirely theoretical.

Data from a nuclear test is gathered in picoseconds. All of the hundreds and hundreds of U.S. tests add up to less than a millisecond of information. Some tests were designed to measure effects, but no tests can gauge the behavior of multiple weapons in a real world setting.

"We have always given a lot of thought to how a nuclear weapon would behave in war: How vulnerable is it? How lethal is it? What are the secondary effects, like the fact that radar can't see through a fireball? You just can't know what works unless a war starts."

Larry was squarely middle-aged, with dark hair beginning to retreat, aviator glasses, and a beard that came to a point; there was a glint in his eye; and he delivered his remarks like punchlines.

For a dozen years, Larry worked on detonation physics and thermonuclear weapons design. Then he moved into management. His current title was Chief Strategist. In the past, I had asked exactly what that meant. There was a good deal of bureaucratic arcana involved, but Larry apparently was being paid to peer into the future. He made recommendations for long-term planning, meaning where technology was headed and, therefore, where the money should go five years down the line. "We would consider cutting back our investment in computers," he said, "or, starting up a development program for a weapon even if there was no specific order for it from the military."

Larry Madsen, like Del Bergen, was a midwestern farm boy. He attended college at the South Dakota School of Mines and Technology. As a graduate student at Iowa State, he went into nuclear engineering and earned a Ph.D. A fellowship from the Atomic Energy Commission obligated him to work five years in the nuclear field. The lab recruited him right out of school, in 1968, "when nuclear energy looked like the wave of the future."

Back then the lab was far more loosely structured and seat-of-the-pants than it would be later. "The physicists and chemists and engineers worked side by side. Whoever drew the short straw was the manager. There was no money or honor in being a manager, you just got stuck with the paperwork," he said. "When it was the Pony Express, even the Post Office was exciting to work for."

Larry did his weapons work when designers could get under the hood of a bomb and fiddle with the parts. "We got an idea, then tracked down whatever new physics or strange materials we needed, making it up as we went along," he recalled. "For instance, when we found out Coors Brewing up in Colorado had a first-class ceramics capability, we made it our business to get hold of some for a test device. At the test site, we used to sit at the wellhead, doing assembly and wiring. There was some last-minute decision-making—you'd take out a part and file it down to size. It was pretty casual."

Part of Larry's new job was playing war games. War games were as close as anyone ever got to a nuclear battleground.

"Every summer, a thousand or so people go to the Naval War College in Newport, Rhode Island, for global war games," Larry said. "They bring in economists, soldiers, politicians, and scientists to work through minute-by-minute decision-making in a set of complex situations. In a recent game, we studied a scenario in which the Ukraine had dropped a nuclear weapon on Russia. The question was, what do *we* do on the day after?

"You are divided into cells and subcells. There is a Blue president, meaning a Western leader and a Red president on the other side. There are cells that represent organizations like Congress and NATO, and smaller groups that do things like deciding when to convert General Motors over to building tanks.

"During the Cold War era, I participated in a war game that went on for five years. Each summer we came back and played out a conventional war between East and West in central Europe, twenty war-game days at a time. The CIA played the Russians. The War College staff acted as controllers and referees. They would assess the effects of your decisions and throw in monkey wrenches, to simulate the uncertainties of a real war. I remember one commander who plotted out an elaborate airlift of reinforcements into Germany. The referee ruled that two of his planes crashed in midair in the fog over London and the counteroffensive failed. The guy was madder than hell and tried to complain, but the lesson was, he had planned too close to the edge. That is the kind of thing you learn."

During the five-year-long game, Larry was part of the nuclear cell. "We found that nuclear weapons were not particularly useful. We did not retaliate with nuclear weapons when we were attacked with chemical weapons. When we were losing, there was no utility in lobbing them over our shoulders as we retreated. If we were winning, there was certainly no need for them. The only time they were seriously considered was in a stalemate. It was proposed that nuclear weapons be used to help take back some lost ground and force peace talks. The British air marshal playing the part of Supreme Allied Commander for Europe argued that this was historically NATO policy, to use the weapons to create a firebreak to give us time for diplomatic recourse. That was the nuclear-war-fighting strategy, that was flexible response, to shoot, look, negotiate, and shoot again if necessary.

"What was disconcerting was that the military players had no idea how to use them or how many they needed to win a battle."

Theories about nuclear war inhabited think tanks and college classrooms. What was tangible were the warheads mounted on the wings of fighter-bombers and in the bellies of B-52s and the nuclear-tipped Trident missiles that could be launched from a submarine and strike within a few hundred yards of a target five thousand miles away. The real nuclear weapons were under the command of warriors, not academicians.

Theory had evolved to say that a multitude of accurate weapons, small and large, short- and long-range, would be the most credible and so the most effective nuclear deterrent. The theory did not say how many was enough.

The theorists did not buy bombs; military commanders did. Generals decided how many bombs they needed to destroy a tank battalion or a bridge. The Pentagon drew up a single list of targets for potential nuclear strikes. At the top of the list were a couple of thousand of the Soviets' own nuclear weapons installations. Farther down the list were targets like a soldiers' barracks in Kazakhstan, and a factory that made fertilizer. The idea was that Russians who survived a nuclear exchange would not have the fertilizer to grow wheat and feed themselves.

There were rules of thumb for nuclear battle plans:

Out of one hundred missiles launched, seventy-three will arrive with 95 percent certainty at a particular point. Seventy-two of those will detonate with 95 percent certainty. To *guarantee* destruction of just one target, a number of weapons had to be launched, separated in space and time so they would not blow each other up. For a thousand targets, get three thousand weapons.

Cleaner, safer, more reliable weapons kept going into production, and the military bought them to augment or replace those already deployed like a noose around the Soviet Union, Eastern Europe, and China. Every nuclear step the Soviets took, the United States and its allies met with a countermove.

"The lab created temptations," Larry said. "Congress would say no to a missile and the scientists would say, 'Well, we can make it smaller and cheaper.' Sometimes I said to myself, 'We're hemorrhaging here.' And sometimes we scared ourselves. Intelligence would see a new Soviet weapon and predict a massive buildup and in the end the Russians only built three."

Ten thousand targets. More than one weapon per target. A steady supply of new, improved weapons. A demand for item-by-item equality with the enemy. That was the arms race; that was overkill.

Nuclear deterrence theory said that weapons must be completely convincing to stop a war. With intentions that were somehow the same and yet not quite, the military wanted nuclear arms that could fight a war when the command to launch came over the wire. Generals wanted enough firepower to win. This was true despite the fact that no one knew what fighting or winning a nuclear battle actually consisted of.

The attachment to nuclear weapons did not always run deep. "The Army looked at the nuclear weapons in Europe and thought, 'We can't play with these toys,' " Larry said. "A lot of officers thought nuclear responsibilities were too tough on a career where a single mistake on your record could scuttle your future." With nukes, there were so many rules and safeguards, it was impossible to pass an inspection. Even something as trivial as improperly inflated tires on a nuclear weapons transport truck earned a black mark. "Out in the field, the weapons were pariahs," Larry said.

Despite that disaffection, the strategy of deterrence triggered an arms race and that arms race mimicked war's messiness, miscalculation, overreaction, miscommunication, imprecision, and misjudgment of the enemy. Whatever the grand plan, someone was in the trenches making moment-to-moment decisions. How many was enough? It was a question that had to be answered with national survival as the stakes.

Larry liked to compare the practice of deterrence with a style of chess called Kriegsspiel. "Two players with two separate boards sat back to back. White had to play the game without knowing where Black's pieces stood and vice versa. A referee said yes or no, that is an acceptable move, and the players used that imperfect information to make deductions about what to do next. Of course, it was easy for someone on the sidelines to say, 'Dumb move,' but for the player, there was no way to figure that out in advance."

Jas Mercer-Smith had made the same point to me a different way. He refused to second-guess the Cold War arms race. "To criticize the arms race is the worst kind of Monday morning quarterbacking. It is laying blame without giving the leadership credit for getting it right—after all, we did not blow ourselves up. Each decision made seemed right at the time, and the penalty for failure was extreme."

•

"WE HAD THOUSANDS OF TACTICAL NUCLEAR WEAPONS stashed in Europe and they were useless," Del Bergen said. "They were called tacticals, but they were too powerful to use."

I was at Bergen's house, in the airy living room that aligned with the third green of a golf course. Ducks and golf balls splashed intermittently into the water hazard that lapped at his backyard.

"The idea was to put a fence around communism, to hold the status quo so there would be time to bring about peace through diplomacy. The fence would allow time for the export of self-determination."

Del laced his fingers behind his head and spread his elbows in an attitude of active meditation. "Containment didn't work, of course. There is North Korea, China, Cuba, and a few African nations—nuclear weapons did nothing to keep communism out."

Lacking the manpower to match the huge Soviet forces arrayed in Eastern Europe, NATO resorted to tactical nukes, thousands of them. Tactical nuclear weapons were small, mobile, handy: designed to fight battles on the ground.

The public imagination locked on to the image of a U.S.–Soviet Judgment Day exchange: flotillas of bombers humming past the point of no return; a hailstorm of missiles; concentric circles of hell sketched over Chicago and Washington to demonstrate degrees of death from a one-megaton thermonuclear explosion. That was only one of many nuclear nightmares, and not the most probable one at that.

In 1957, a congressional committee asked the Chairman of the Joint Chiefs of Staff, Admiral Arthur W. Radford, if the use of nuclear weapons should be renounced, as poison gas had been in the past. He answered, "Gas was not used in World War II because it was not decisive. Atomic weapons are going to be decisive. . . . We have integrated them into our plans and we expect to use them in the event of aggression against us."

"We put nukes into Europe as fast as we made them," Del said. Los Alamos designed the Honest John, a long, skinny rocket manageable enough to launch from a rail mounted on a truck trailer. By the early sixties, hundreds of Honest John missiles rolled through the pine forests and pastures of the German heartland. Their nuclear warheads packed twenty kilotons of explosive power, the same as the atomic bomb dropped on Nagasaki. The Army also issued a recoilless rifle, the Davy Crockett, with an atomic projectile so small an average woman could lift it. Davy Crocketts could be hauled anywhere a jeep could go.

It was edgy along the border. "The Soviets had frontline atomic weapons," Del Bergen said. "Their troops were equipped with gas masks. We didn't know what they were up to."

Tens of thousands of American soldiers held forward positions on NATO's central front. Nineteen-year-old boys from South Carolina with M-14 rifles slung over their shoulders kept watch on the Russians a few hundred yards away, patrolling a thin barbed-wire frontier.

One-third of West Germany's population lived within 150 miles of the East German border. What was going to happen to all these Americans and all these Germans if combat in Europe crossed the nuclear threshold?

Weapons designers described German villages as "two kilotons apart." The smallest nuclear weapons were so destructive that even in a limited exchange, military and civilian deaths would mount astronomically. Nothing that could be described as victory could be achieved. Every town NATO was defending would be rubble after a few nuclear weapons were lobbed back and forth. Any troops that were still alive would be stalled in place, unable to advance into territory contaminated with radioactive fallout. Tactical nuclear weapons on trucks dispersed over a wide area would be difficult to control from a central command position, but relatively easy for an enemy to capture.

"At the time, they did not know *how* to make plans for nuclear-war fighting. Field commanders were in charge of the missiles. *Nike* missiles couldn't hit the broad side of a barn," Del said. "If you really look at it, the weapons didn't make sense."

Thousands of tactical nuclear weapons, trapped in this logical fault zone, remained in Europe for the duration of the Cold War. As materiel of war, they were pointless. As deterrents, they did the job in their own backhanded way.

The first time in history that Americans and Soviets pointed guns at one another was at Checkpoint Charlie, a gateway through the Berlin Wall, in October of 1961. It was one of the final acts in a war of wills between East and West. No shots were fired.

Earlier in the year, Khrushchev had published a plan to annex West Berlin and, he said, "If in reply the imperialists unleash a war, hundreds of millions might be killed." It was clearly nuclear war he was talking about, and President Kennedy answered in the same language. Kennedy asked Congress for authority to triple the draft, and for billions of dollars to strengthen the armed forces. In a televised speech, Kennedy announced a $200-million budget request to identify fallout shelters and to stock them with food, water, and first-aid kits.

The United States and NATO did not have the troops, tanks, or artillery to defend West Berlin, and nuclear engagement was given some weight as an option. Yet, every possible plan, from a nuclear warning shot to a preemptive strike on the Soviet Union, was rejected out of hand.

In the midst of the crisis, Khrushchev built the Berlin Wall, which put an end to the problem of escaping East Germans. Kennedy met the move with silence. Two months later, tanks rolled into Checkpoint Charlie in a confrontation over the right of Western personnel to pass freely into the Eastern sector. The face-off lasted a few days before Khrushchev backed off. Khrushchev left West Berlin alone; Kennedy accepted the Wall; there was no war.

There were hundreds of atomic weapons along the front, but calculations of tactical missile ranges and bomber strength had no direct effect on the decisions that were made: in fact, the United States had overwhelming nuclear superiority. It did not matter. Even the smallest risk of nuclear retaliation was too great.

"The nuclear weapons forced the United States and the Soviet Union to talk," Del said. "After Berlin, and after the Cuban Missile Crisis a year later, the hot line was installed."

Nuclear weapons were like a capricious god that nobody wanted to rouse. Larry Madsen had said, "We are a nation that argues over the rights of laboratory rats. How could anyone think we would bomb Germany?" But Jas saw it differently. "Part of deterrence is that we never know if we will use them," he said. "After all, we have done it once." Deterrence found its footing on a razor's edge of uncertainty.

It is slippery to characterize events that did not

happen, but it is fair to argue that nuclear deterrence prevented direct gunfire between the United States and the Soviet Union. Nuclear weapons deterred the Soviets and they deterred us. People died in combat anyway. Having a supply of nuclear weapons did not keep a nation out of war altogether: the United States fought in Korea and Vietnam; the Soviet Union fought in Afghanistan; Great Britain went to war over the Falkland Islands. Apparently, little countries did not fear that bigger ones would bomb them with nuclear weapons. Apparently, big powers found no good reason to use a nuclear weapon, even when they had no fear of retaliation.

Years ago, there was a particularly enigmatic toy around. It was a nearly seamless black box whose only feature was a toggle switch on one side. When you turned it on, the top of the box opened. A plastic hand came out, reached over, flipped the switch off, and disappeared back inside the box. Similarly, finally, a Cold War nuclear weapon's sole function was to turn itself off.

"WHAT WE HAVE TO DECIDE AS THE COLD WAR ENDS IS, why do we need nuclear weapons at all? If we need them, the question is how many and for what purpose?"

John Hopkins was at the Institute on Global Conflict and Cooperation (IGCC). The offices had a blue-sky view of the University of California at San Diego where the University met the Pacific Ocean. We were close enough to feel the mist off the waves. The University's cool lawns shaded by eucalyptus trees were a soft landing place for someone used to Los Alamos's tough desert hide.

The Institute was a think tank for research into nuclear arms and other international security issues. It was the brainchild of Herbert York, a former director of the Livermore weapons lab and presidential adviser.

Hopkins was on loan from the lab, putting in a year at the Institute. At one time, Hopkins was Test Director at the nuclear proving grounds in Nevada. He later participated in arms control talks in Geneva—he described his role there as "token hawk." John was in his late fifties, with bright eyes and a tidy gray beard. He was an easygoing, animated man.

"I used to believe that *somebody* in the government must be thinking about the long-term implications of nuclear weapons. As I moved through the ranks, I kept figuring the guys must be on the next higher level. In fact, no one in the Department of Defense or the Department of Energy does that kind of thinking. Not many working stiffs read *Foreign Affairs*. Along the line, there have been a few men who have addressed the larger issues—Nitze, Schlesinger, Brown, Kissinger—but for the most part that kind of thinking has only been done in places like the Institute."

Perhaps the end of the Cold War was an opportunity for the doers to fall into step with the thinkers, even thinkers so far away from Washington they could hear the surf drumming in off the Pacific.

"What do we do with nuclear weapons now?" John asked. "The only plausible role for U.S. nuclear weapons is last-ditch self-defense, guarding our borders. It is extremely difficult to identify any other credible use for America's nuclear weapons. They are simply too indiscriminate and destructive to use in a regional conflict."

Hopkins foresaw U.S. nuclear weapons shrinking in significance down to a single pinpoint policy. But our nuclear troubles were not going away—they were convoluting and slipping out of our control. They were relocating.

"The end of the Cold War eliminates the threat of instantaneous destruction. But now the chance of Third

World countries exchanging nukes becomes very likely," John said.

The small but real risk that we would *all* die in a nuclear holocaust had been replaced with a greater risk that *some* of us would die by the bomb. It was a familiar anomaly of nuclear arithmetic: "In large numbers, the bombs make major war impossible. In small numbers, they can be very effective. If Iraq had had two in its war with Iran, that would have made the difference. Nuclear weapons have war-fighting capability in an uneven confrontation."

Apparently, its nuclear forces cannot provide the United States or any allies with much protection against threats from renegade states in the post–Cold War world.

"Eventually, the U.S. will confront a nuclear-armed developing country. What then? It certainly limits our options. Right now, it is hard to imagine a nuclear response in that situation, though this may not always be the case. Frankly, I can't imagine a superpower without nuclear weapons, or a superpower using them."

The greatly reduced utility of nuclear weapons presumably was going to affect the size of the stockpile.

"For self-defense, they are desperation weapons," Hopkins said. "All we need is a stockpile in the high hundreds or low thousands."

What was the meaning of those numbers? A few thousand was still enough for Armageddon. With a stockpile of a few hundred nuclear weapons, targets would once again be cities, instead of military installations.

Jas Mercer-Smith had said, "When you have too many nuclear weapons, mistakes, accidents, and theft are more likely. Too few, and you introduce different problems. The two hundredth-largest U.S. city is a town of fifty thousand, so two hundred of someone else's weapons could destroy *us*. We could do the same to them. If you go down to only a hundred, then each one becomes too valuable. You would be tempted to use them early on in a conflict, rather than worry about their vulnerablity." Use them or lose them, in other words.

Larry Madsen had his own approach to the question. "There is the blob theory, which is that we don't need to rationalize an exact number—that we need some, but it is targetless deterrence," he said. "I personally could go to zero if I felt secure that my conventional defensive and offensive weapons could threaten sufficiently for deterrence. Keep in mind that conventional weapons are really grisly."

There was plenty of evidence that the United States was ready to wash its hands of old-style nuclear doctrine and to drastically reduce the stockpile.

The Strategic Air Command quit operations in June of 1992. Their B-52 bombers no longer endlessly circled near Russian airspace. Every last European tactical nuclear weapon was scheduled to be either withdrawn from Europe or destroyed. South Dakota silos had housed Minuteman intercontinental ballistic missiles for thirty years. The Air Force emptied them out and plugged them up. The United States Army and the United States Marine Corps handed in their warheads, leaving all nuclear business to the Air Force and the Navy. A moratorium brought nuclear weapons testing to an end in 1992. The nuclear stockpile, at around 20,000 warheads in 1990, shrank to 16,750 in 1993, on its way down as fast as the Pantex workers could manage it. Processing plants that had manufactured uranium, plutonium, and tritium for nuclear weapons were closed down.

In 1994, the Department of Defense undertook what it called the Counterproliferation Initiative that will equip U.S. forces with nonnuclear technology to battle an enemy armed with "special" (nuclear, biological, or chemical) weapons. It is the answer to the question, What if Saddam Hussein had a nuclear bomb? Another possible answer—one that few were eager to adopt—was to design mininukes, low-yield, high-precision weapons that could, for instance, blow up a biological weapons stockpile without dispersing the pathogens.

For the time being, the traditional nuclear stockpile was going down, and no new weapons were being designed or built. The usual job of a laboratory like Los Alamos had to be eliminated or revised. The nuclear technowizards were looking around for different ways to apply their secret skills.

Nuclear strategists were not, however, altogether ready to nail the door shut on the weapons business. The lab continued to have statutory responsibility to keep a cadre of designers, and there was a standing presidential order to maintain the capability to return to underground nuclear testing on six months' notice.

"A lot of what's happened is simply because of budget cuts," John Hopkins said. "There is still a policy vacuum."

Weapons designers went into a holding pattern. The mammoth installation at the Nevada Test Site was in suspended animation. Just in case.

Herbert York, the founder of the IGCC think tank, was an old man with a stern brow and a gentle temperament. The prototypical defense think tank, Rand, had been established by the Air Force. York wanted the IGCC to draw on the intellectual capital of a university.

He lived a few miles south of the Institute in a house

footed in the cliffs at the edge of the ocean. His balcony overhung a smooth rock ledge that shimmered with tidal pools. He stood there watching gulls skim over the Pacific. "Ten years from now," he said, "the important thinking about nuclear strategy will take place in Jerusalem and Baghdad."

CHAPTER

3

OUTSIDE LAS VEGAS

> Technically, we learned something very important from Hiroshima and Nagasaki. These things *want* to go off. We used two different kinds of bombs and they did not resist exploding. Then the problem becomes *keeping* them from going off.
>
> —Bob Brownlee,
> Los Alamos National Laboratory

FROM THE CAR, THE DESERT LOOKED FLAT. AERIAL photographs show hundreds of broad, smooth dimples on the surface of the Mojave, but from the car, nothing.

Don Collins was at the wheel. He was saying that the Nevada Test Site is as large as Rhode Island. I had no idea that Rhode Island was so small, but I kept my thoughts to myself. Don had worked in nuclear-weapons testing as a general factotum since 1956. He met his wife, a nurse, at the dispensary in Mercury, the little slapdash town that serviced the test facilities. He had seen nuclear weapons dropped from airplanes, dangled from balloons, and shot from cannons. His pride in this enterprise was airtight.

The pockmarks in the desert were subsidence craters, wounds left behind by underground nuclear explosions. At

Frenchman's Flat, I saw more instructive ruins: the twisted steel and scorched concrete that memorialized the atmospheric tests of the fifties. At one time there was an effort to cart away all the blasted debris from Frenchman's Flat, but the Atomic Energy Commission decided to let it stand as evidence.

"You can't build anything or tear anything down without permission from the historic review board. Or, for that matter, from the group that protects the desert tortoise habitat," Don said.

The soil at Frenchman's Flat was as fine as talc. It hardened into greasy, gray scales like crocodile skin. The condition was entirely natural and unrelated to any warcraft. The bridge that had snapped like a broken pencil was man-made, and so was the eight-foot aluminum dome, crushed and folded like a discarded beer can. The ground was littered with cables, steel beams half buried in the clay, and bits of wire.

Inside the remains of a small building, a bank-vault door stood intact, but the surrounding concrete walls had been peeled away by the force of an atomic blast, leaving a skeletal forest of rebar bent back like trees in a hurricane.

Farther along was a concrete dome, apparently untouched, while its nearby twin was a pile of rubble. The smashed dome had six-inch walls before a nuclear bomb went off a thousand feet away. The survivor's walls were two feet thick.

"The dome is still standing, but it got up to eight hundred degrees Fahrenheit inside—it was no place to hide," Don said.

To measure the effects of atom bombs, the Army used to fabricate wooden houses, brick houses, and canvas field hospitals; they planted airplanes and school buses in the path of the explosions. They used pigs to test biological effects, Don Collins told me.

"The pigs wore tailored uniforms," he said. "We had a working field hospital in tents a safe distance from the tests. Any pigs that survived would be taken there. The idea was to practice medical techniques where radioactive fallout was a factor."

Now at Frenchman's Flat, anything that could go up in flames was long gone. The tortured scraps of iron and stone that endured were dormant. They divulged no more about the moment of nuclear detonation than ashes speak of the heat and color of fire.

DON COLLINS WAS IN THE ARMY WHEN HE WAS ASSIGNED to the Nevada Test Site to run the motor pool. He left the service but stayed on in Nevada, keeping his job by changing his employer to Los Alamos National Laboratory. His responsibilities expanded into all kinds of housekeeping duties like visitor protocol, shipping, communications, and generally seeing to the needs of Los Alamos personnel who came to Mercury. Now he was in his late fifties, but tireless and very fit, an ageless package of enthusiasm, a born salesman.

It was 1991, and nuclear testing was still business as usual in Nevada. Collins had pulled in front of Caesars Palace at 7 A.M. to pick me up for my official visit. The morning sun bleached the glitter right out of the Las Vegas Strip—the town almost looked ordinary as we motored out of the neon canyon and into the naked desert. We made one stop to pick up Karen Randolph. Karen was a public relations officer for the Department of Energy, but it was Don who did all the talking. Don talked a wide blue streak.

As we drove north and west on Highway 93, he told a

story about a fatal avalanche in the Spring Mountains, a small range between us and the California state line. We passed through a few small towns and Don reminisced about earlier days, when fistfights would break out in the bars where the Air Force boys came to drink after dark.

"Nellis Air Force Base and its gunnery range surround the test site on three sides—great built-in security," Don said. All there was to see were empty tracts of sandy earth so flat and raw they might have been scraped out by a bulldozer.

The town of Mercury sat near the entrance to the site, just inside a high chain-link fence—closed to the outside world. We swung through Mercury, taking in the bowling alley, the two-story dormitories, the trailer park, the warehouses, the swimming pool, and Don's office. "This building used to be the town's all-denominational chapel," he said. "Recycled it."

Don's chronicle continued as we headed north into the teeth of the tour. "Paiute—they were hunter-gatherers—used to crisscross this area. These are dry lake beds. You can find springs about every five to ten miles."

Later: "We used to pay for dinner at the cafeteria by slipping a silver dollar into the turnstile."

At a rise overlooking Frenchman's Flat: "We set up bleachers right here for the press to watch tests go off—Edward R. Murrow came once."

We drove all day and Don never ran out of anecdotes.

"The Russians once put out a gas well fire with an underground nuke."

Karen Randolph, still with us, listened just as hard as I did. "I never knew that," she would say from the backseat.

Around midday, I came up with a theory about why

Karen was along for the ride. Apparently, the rules called for me to be escorted the *entire* time I was on site. Karen was there to follow me into the ladies' room.

We bought lunch in a windowless employee cafeteria. Crews of roughnecks and teamsters, secretaries and mechanics ate chicken-fried steak in a yellow fog of fluorescent light and cigarette smoke. A vending machine dispensed beer for a dollar.

Don wanted to be sure I caught on to the scenery. Back in the car, we detoured to visit a pond, a dug-out well, where coots and kingfishers paddled among the reeds.

"We have bobcat. We have antelope. Coyotes come up and try to steal your lunch. Up in the mesas, it's beautiful. There's mountain lions up there." This was not propaganda. Don loved the desert. Later, we watched four wild horses, sturdy, high-headed mares, come down out of the hills and drink from a holding tank in the noonday sun.

NORTH OF FRENCHMAN'S FLAT WAS YUCCA FLAT. THE road was a straightaway through low clumps of blackbrush and creosote.

"The most dangerous thing out here is the traffic," Don said. "The monotony of these roads is like a drug. We're picking bodies out of wreckage all the time."

The desert—broad, searing, and alkaline—overwhelmed the testing operations. Industrial cranes a dozen stories tall looked like saplings against this landscape; a haze of dun-colored dust camouflaged the buildings.

There were close to seven thousand employees at the site, but the action was spread pretty thin over the 1,350 square miles. The day I took the tour, it so happened that the nuclear device for a test code-named "Lubbock" was going into the ground. Thirty or so construction workers, a crowd in these parts, were easing cables down the drill hole. They stood shoulder to shoulder beneath a raised steel platform where scores of electrical lines were suspended and fed out ever so slowly. In a relay of smooth moves, passed from one gloved hand, one bare, muscled arm, to the next, the men guided the lines into a hole hardly more than six feet in diameter. It was a tight fit.

These cables were a thing unto themselves, the arteries of the experiment. They served to carry the firing signal down into the hole, and then bring back all the data that could be registered in the instant before they were consumed by a nuclear blast.

"They have to be wrapped in white plastic—the sun would just murder the standard black we used to use," Don told me. "The test rack goes down sometimes as far as two thousand feet. The cables—they're coaxial and optical fiber—can't stretch or twist or the signal degrades. Each line has to have a block to keep radioactive gas from shooting back up out of the hole. This cable costs between five and ten dollars a foot. You can have several million dollars' worth of cable on a single shot."

I stood directly on top of a thermonuclear explosive, trying to measure my emotions. The hardhats were lowering a multimegaton hydrogen bomb into the shaft. The work going on around me was steady but not urgent. Some of the men, having taken cover in whatever strips of shade they could find, were watching and smoking. I waited for the hair to stand up on the back of my neck. I tried to picture a second sun in the earth below my feet. It should have been easy—the true sun was right there to look at. Even though it is colder than a fusion bomb, the sun still hits the Mojave so hard it can blister skin from ninety-three million miles away.

Don Collins did not need his imagination. He had seen scores of nuclear devices exploded in the air. He remembered "Priscilla" this way: "I was seven point one miles from ground zero. It was a balloon shot. There's the flash and then the first sensation of heat—hot, like a sunlamp, but not hurting. No noise yet. After the fireball goes up, you see the shock wave of dust racing across the desert. The sound got there with the dust."

"Priscilla" would have been a low-yield shot, the test of a fission device meant to be a component—the starter motor—for a fusion weapon. The biggest bombs were tested in the Pacific. A sailor who watched the "Mike" shot, the world's first thermonuclear device, said, "It would take at least ten suns to equal the light of the explosion from a distance of thirty-five miles." Another wrote, "A flame about two miles wide was shooting five miles into the air. This lasted for about 7 seconds. Then we saw thousands of tons of earth being thrown straight into the sky." From another eyewitness: "You would swear that the whole world was on fire."

When a nuclear bomb explodes above ground, its unearthly energy collides with the surrounding air and, in less than one millionth of a second, ignites a sphere of fire, fiercely incandescent, that shoots upward, expanding at supersonic speed as it goes. The drag of the air begins to cool the outer surface of the fireball, but inside, the gases are still raging hot. The hotter gases rise faster, creating a roiling doughnut shape.

Del Bergen is another veteran of atmospheric testing. "As it lifts, the fireball draws loosened ground up into the vacuum at the center of the doughnut. That is the stem of the mushroom cloud: a column of debris that swiftly connects to the bubble.

OUTSIDE LAS VEGAS

"I watched a hundred kilotons from fifteen or twenty miles away. (You can see a ship on the horizon at fifteen miles.) The light was horrible, even through welder's goggles. You can see features in the light—the flash is so fast your iris is still open. Then comes the shock wave," Del said.

Energy is like a substrate, the ether out of which everything else is fabricated. Matter itself is energy translated into quiescence. A nuclear detonation pries open the locks that hold the energy of fundamental matter in repose and for the briefest flicker of time concentrates it in a small space, a transcendental eye of chaos. A nuclear explosion is what happens when this dense convergence of energy rushes outward in its urgency to cool, to settle, to find equilibrium with the ordinary world.

In the first millionth of a second, a bomb's energy radiates out as X rays, but the air surrounding an explosion instantly absorbs the X rays and moderates the energy into heat and light and shock.

The blast from a nuclear explosion crashes through air, moving several times faster than the speed of sound. Here is James Mercer-Smith's description: "The shock wave heats the air a lot and quickly. It moves faster than the air molecules can transmit information, so you get a discontinuity with a sharp edge, an abruptness. It is not so much the high pressure that hurts; the force comes from the incredible *change* in pressure.

"An air-pressure difference of only five to ten percent is enough to cause a hurricane-force wind. The amount of energy in a nuclear explosion is less than an actual hurricane—the pressure difference begins in a space a few meters across as opposed to hundreds of miles across—but a bomb creates a trillion percent pressure difference."

The heat and light pulsing out from a nuclear fireball are roughly in the same range as the sun's radiation. The heat will vaporize everything within a few hundred meters of an explosion. The visible light is so intense it strikes like a hammer.

A standard 1,000-megawatt coal-burning electric power plant routinely generates one kiloton of energy—about one-tenth the yield of the Hiroshima bomb—but the power plant does it over the course of an hour. The bomb does its work in nanoseconds. A nuclear bomb creates temperatures measured in tens of millions of degrees. As in the case of the shock wave, the amount of heat by itself is catastrophic, but added to that is its awful suddenness.

Inside the fiery extremes of detonation, a fraction of the energy transfigures itself into ionizing radiation: gamma rays and neutrons that constitute a kind of invisible subatomic shrapnel. This radiation shoots out, penetrates most ordinary material, and travels great distances. Like a silent bullet, ionizing radiation tears up the molecules in a body's living cells.

The shock wave, the heat, the light, and this fusillade of radiation spread out and dissipate fairly quickly. A final small portion of a nuclear explosion's energy lingers, bound up in the atomic dust that is all that remains of the body of the bomb. This is fallout.

Uranium and plutonium atoms from the bomb, shattered by fission, break into a constellation of new-minted elements. These elements are born in a state of imbalance and internal agitation. As the new atoms stabilize, they emit their own variety of violent, ionizing radiation, but piecemeal, over hours, days, years. These radioactive elements attach themselves to soil and water, and any other material that has been drawn up into the maelstrom of the fireball. Fallout rains back down to earth where it can be blown around in the wind, eaten, breathed in, and touched.

"The amount of damage a one-megaton bomb can do—you can't absorb it," Del said. I assumed that, as always, his words were picked for precision. You cannot grasp the force of a nuclear explosion by multiplying normal experience. You can put numbers on the event, but they spiral upward into meaninglessness: a million times normal air pressure, tens of millions degrees of temperature, phenomena that are too big for the planet.

The last aboveground U.S. nuclear test, "Tightrope," went off fifteen miles over Johnston Island in the Pacific on November 6, 1962. France continued atmospheric testing another dozen years and China went on until 1980. Since then, atomic explosions have disappeared from view.

In the fifties, both U.S. and Soviet scientists had been detonating nuclear devices at a machine-gun pace—there were well over a hundred tests in 1958.

At the same time, many scientists, most notably the Nobel Prize–winning biochemist Linus Pauling and Andrey Sakharov in the Soviet Union, warned against long-term damage from radioactive fallout. Bomb debris from atmospheric tests, spread far and wide by winds in the stratosphere, was likely to cause cancer and birth defects in people residing on either side of the Iron Curtain. Strontium 90 was sinking into the bones of growing children and polluting the milk they drank.

Khrushchev and Eisenhower agreed informally to suspend all nuclear-weapons testing after 1958. Anxieties about fallout nudged them along, but in fact they had other reasons to negotiate an end to all testing forevermore. A test ban would stop the extravagant, hand-over-fist buildup of nuclear arms. Without testing, weapons technology—if not stockpile numbers—would simply freeze at the current level.

While the test sites lay quiet, negotiators two-stepped around a treaty for a couple of years. It is safe to say that both sides genuinely wanted to stop the arms race. The problem was to define the amount of information that needed to be exchanged to make a test ban work. In the matter of information, Khrushchev and Eisenhower were in impossible positions. American suspicions demanded absolute reassurance that the Russians could not cheat without getting caught. Even though there was general agreement that it was technologically possible to know of any atmospheric or underground tests that might violate a treaty, advisers to Eisenhower kept spinning out frightening scenarios, such as Soviet nuclear tests in deep space that would be undetectable. Khrushchev's problem was entirely different. Years after the fact, Khrushchev wrote that he could never have agreed to test-ban inspectors probing deep into his territory: they would have discovered the Soviet Union's relatively weak position, and this, he feared, would encourage a nuclear first strike from the West.

On May 1, 1960, Francis Gary Powers's U-2 spy plane was shot down over Russia. Khrushchev was furious and Eisenhower was unapologetic. Thus ended the fragile trust upon which the test ban talks were predicated. Eisenhower left office with nothing to show for his efforts at arms control.

The next year, 1961, was a year of fearsome confrontation. In April, newly-elected President Kennedy allowed the disastrous invasion of Cuba at the Bay of Pigs. In August, Khrushchev built the Berlin Wall overnight. The caution that characterized the moratorium had been replaced with combativeness. In September, National Security Advisor McGeorge Bundy woke Kennedy from an afternoon nap to report that the U.S.S.R. had just fired off the first nuclear test to be staged in three years. The Russians banged out three more in the space of a few days. The United States responded in kind: within two weeks Los Alamos scientists fired a test of their own.

Del Bergen was in the thick of things at the time. "Resumption of testing was a political statement," he said. "We rushed in and those early tests were not as well planned or productive."

In 1962, nuclear devices went off like strings of firecrackers: between us and the Soviets, more than one every three days. In the fall of that year, the duel over Russian missile sites on Cuban soil allowed both Kennedy and Khrushchev to peer directly into the nuclear abyss. What they saw sent them back to the negotiating table for another try at a comprehensive test ban. They fell short of that, but by August, the Limited Test Ban Treaty was hammered together and signed by the United States, Great Britain, and the Soviet Union. The treaty banned testing in the atmosphere, in outer space, and underwater; it went into effect October 10, 1963, a little more than a month before Kennedy was assassinated.

Announcing the treaty on television that fall, Kennedy called it, "a step towards peace, a step towards reason." Perhaps. Even though environmental and health concerns figured very little in the test-ban talks, the environment was the only beneficiary of the treaty. It would be hard to argue that the limited test ban made a dime's worth of difference in the arms race.

During the treaty talks, weaponeers at Los Alamos were regularly called on to counsel the administration on the subject of nuclear testing. Some of them were relentlessly opposed to curbs on testing; some of them were not. Bergen saw it from the inside.

"In the fifties and sixties scientists had an almost cult status, like star athletes today. The press would ask rocket scientists questions in fields they had no business talking about. I think it's true that Los Alamos scientists had influence larger than they intended, larger than they wanted, and larger than they deserved.

"At the time, most Los Alamos weapons designers were mildly anti–underground testing," Del recalled. "The more we learned, the more we understood the magnitude of the fallout problem, but we were slow coming to it."

The designers' reluctance notwithstanding, Norris Bradbury, director of Los Alamos National Laboratory during both the Eisenhower and the Kennedy administrations, had more than once testified in front of Congress that he could live with a test ban.

Del worked under Bradbury for years. "Bradbury's attitude was, if you want us to go do something, we will do it—let's get this job done and go out of business. In that way he was like Oppenheimer."

After twenty years of blowing up their devices in plain sight, the designers were now faced with the problem of extracting all their information from under tons of earth.

Del said, "We didn't like the idea of going underground. Obviously, it was going to be hard to see what was happening. And the physics was different. Fast neutrons would fly out of the primary, hit the dirt, and bounce back instead of just sailing into the air. But, as it turned out, it was actually easier to instrument downhole."

Still, if the job is to understand nuclear weapons, underground tests impose a natural limitation. "Effects against targets: that's what we lose because of underground testing. We lose the environmental realities," Del said. "In the last series of tests in the Pacific, we were awed by the possibilities of what we were seeing. High-altitude detonations created a bubble of conductivity that shoved the earth's magnetic field lines out of the way. It was a ripple effect, like an accordion, that generated current, a surge strong enough to shut down a power grid."

No one argues for atmospheric testing—in the light of current understanding, the hazards are clearly unacceptable. Del said, "I was one generation behind the guys that really screwed up—long-term exposure seems to have cut some lives short, especially cyclotron workers. But I used to take a beat-up old chunk of uranium 238 and hit it with a hacksaw just to watch the sparks fly—I don't believe I'd do that now. At the time, we focused on single exposures to radiation, but there was no recognition of the cumulative effect."

Hundreds of nuclear devices have gone off in Nevada in the three decades since the Limited Test Ban Treaty, all of them quarantined inside the earth, but people who have seen a detonation with their own eyes have a special sense of what a nuclear bomb is. Nuclear bombs going off completely contained begin to look manageable.

Harold Agnew, director of Los Alamos National Laboratory during the seventies, a man who had flown on the mission over Hiroshima, said once, ". . . I firmly believe that if every five years the world's major political leaders were required to witness the in-air detonation of a multimegaton warhead, progress on meaningful arms control measures would be speeded up appreciably."

His remark suggests that something besides data can be lost when testing goes underground: a sufficiency of fear.

The "Lubbock" test happened to be James Mercer-Smith's baby. When I spoke to him a few weeks later he reported with pleasure, "When it went off it registered 4.8 on the Richter scale, the third-largest quake recorded in the world that week."

Designers always went to admire the holes they made in the desert, even though the size of their crater depended on the geology around an explosion and not necessarily the yield of the device. Jas's first crater, years before, was huge and sank straight down. "I know it's sophomoric but I loved that hole. It said to me, 'Nature agreed with you.'

"Keep in mind that we do blow it on occasion. We test something and it doesn't work—spectacularly. The diagnostics people look over and say, 'Can I turn it off now?'

"In fact, you have to be an egomaniac to lead five hundred people to test your crank idea. You get one try and it costs something like thirty million dollars."

The nature of the "Lubbock" test was classified. Jas could only tell me that it was "announced." "When a test is going to yield between twenty and one hundred fifty kilotons, it might move the earth a bit, so the public is warned. Window washers on tall buildings in Las Vegas know to strap themselves on a little tighter. Nothing has ever happened, but just in case."

Different tests in Nevada had different purposes. When a specific new weapon ordered by the military was in development, its design was tested at least half a dozen times as the engineering was refined. Once a new weapon design graduated from prototype to mass production, it was tested again to verify that it performed as expected.

A few tests were meant to find out what life in the real world could do to a nuclear warhead. Assuring me that they were speaking hypothetically, both Jas and Del described such tests.

"Let's say you want to know how a warhead would behave if it went off with its cruise missile still half full of fuel," Del Bergen said. "Propellants are hydrogenous. Hydrogen slows and scatters neutrons, so you might look at the asymmetry of having propellant on just one side of the physics package."

"Weaponization," Jas said, "turns an interesting physics experiment into a missile that is made to go eighteen G's and thousands of miles and is built to last at least twenty years."

Nuclear weapons were bounced along the road in trucks. They were flown at high altitudes, going from one temperature extreme to another; they were plunged to the bottom of the ocean in submarines; they could be caught in electric storms or dented by a forklift. Environmental assaults were anticipated in the engineering of the weapon, but testing was the ultimate guarantor of safety.

In language even more circumspect than Del's, Jas said, "It would be possible to test for any changes that transporting a weapon might bring about. You would fly it around, then stick it downhole. That would be a simple test with limited diagnostics."

My guess was that the "Lubbock" test was in another category—it was simply an experiment to try out a new idea. Weapons designers were supposed to set aside conventions and play around. They got paid to dream up new configurations of materials and to invent different ways to thread through the intricate physics of detonation.

"Recently," Jas said, "at least half the purpose of tests have been safety issues, thirty percent for deeper understanding of the physics, and the rest for new ideas."

Making weapons smaller, safer, and easier to deliver was mainstream design. Out on the fringes, there are scientists probing the edge of physics: trying to prod the device into producing certain kinds of emissions, like X rays, or trying to build a bomb with no fallout.

When a fresh idea was mature enough, the designer had to stand up and convince colleagues that it was worth a multimillion-dollar shot.

In traditional science, publishing is the medium of exchange. In the process of publishing the results of an experiment, a scientist's methods and logic are scrutinized by peers whose motive is to ferret out mistakes. To survive the peer review and be published in a prestigious journal defines success. Beyond that, publishing earns the scientist a seat at the table: there will be invitations to international conferences and access to funding. Since the days of the Manhattan Project, the weapons scientists could not publish—even the names of the projects they undertook were classified. Instead, they ran the gauntlet of a few insiders. Rather than a journal article, a full-scale test in the desert of Nevada was the defining event in a designer's career.

Jas enjoyed the ordeal of defending his proposals in front of skeptical coworkers. "Designers are a small, cabalistic group. The dynamic of competition—call it showing off—is critical to the process of creativity and ambition. Hostile peer review is energizing. It makes people push harder. The rivalry between design groups at Los Alamos and Livermore is really juvenile, but it drives us. You use your work to say, 'See how much smarter I am than you are.'

"Just getting a shot isn't enough—it has to be successful," Mercer-Smith said. "Doing it more than once is even more important. In the design group's apprenticeship-journeyman-master system, the hierarchy is based on years of successful testing. After eight years, I'd call myself a senior journeyman, or maybe a junior master. You don't get paid more or get a better office because you have brought off a test. You get the respect of your peers. In our case that's about twenty-five people.

"Enormous resources can be brought to bear to test your high-risk idea, but you have no real authority except persuasion," Jas said. "You have to recruit a whole team. You get a shot, but you have to sign up an experimentalist, then a lead engineer—a lot depends on personalities, because you are going to share management responsibilities with the team leaders. You have to convince a lot of people to do the work the way you want it."

The job of lead designer on a test shot required a special mix of cocksureness and humility. "Other people on a test are doing work just as hard as your own. Even the welding is an art form—a bad weld can be the difference between success and failure. At Los Alamos lots of people 'need to know'—a chemist or a theorist might review what you're doing and be the one to say, 'That's incredibly stupid.' And designers aren't the only ones who come up with breakthrough ideas. An engineer might come to you and say, 'By the way, I can make this metal light as air,' and suddenly you can do something you never thought of before."

The miles of cable that I saw snaking across the sand and into the ground in Nevada were lightning-speed pathways between "Lubbock" and its makers. Down in the hole, along with the nuclear explosive, were about two dozen detectors. Everything was attached to a "rack," an openwork steel cylinder that could be as much as thirteen stories tall. Near the bottom of the rack was the fission-fusion device: the physics package (mock-ups showed it to be about the size of a duffel bag), and tucked in just below that, the timing and firing gear.

It took two to three years from start to finish to organize a test. "I will say I want to measure *this* phenomenon—go design a detector," Jas said. "I have to make a prediction about what should happen and get a diagnostics specialist to make a detector to find it. Even then you build in redundancy—different sensor designs to pinpoint the source of differences between what you expect from computer models and what the actual outcome is. With the proper coverage in diagnostics, you can learn as much or more from failure as from success."

Most diagnostic devices measured the wild outpouring of radiation in the first few billionths of a second after detonation: gamma rays, X rays, and neutrons—their number, their energies, and profiles of their movement through space and time. Others measure the sheer kinetic energy—the physical force of the blast. These measurements raced upward through the cables an instant before the explosion vaporized the wires. A hundred recording stations housed in trailers a few hundred yards from the hole collected the babel of electronic impulses.

"Every detector on a shot is customized for that particular test. The lead designer oversees the development of each one, and the engineering of the nuclear device, constantly making compromises between the ideal and the practical. As critical parts are crafted, the designer goes to, say, Oak Ridge, to watch the manufacture and to sign off on the components."

It was not easy to translate the data from diagnostic instruments back into the reality of what came out of the bomb. An entire group within the design division was devoted to calculating the interaction of subatomic particles with the macroenvironment—wires, cable coverings, hardware—so the final data took into account all possible events between the explosion and the recording devices.

As preparations for the test progressed, the clubbiness of the team was recognized with various trappings, like stickers, hats, and sweatshirts bearing the shot's logo. The "Lubbock" insignia was a cartoon character, Augie Doggie, dressed in cowboy clothes, smiling while bronco-busting an undistinguished horse, branded "FH" on his bucking hindquarters. (Jas did not believe the rumor that "FH," the initials of a well-known theoretician, was meant to suggest that this fellow physicist is a horse's ass.) Cartoon stars of pain shot out from the point where the seat of Augie's pants met the horse's back. The two creatures were framed inside a bulging right triangle, blunted at the corners, along with a legend: "Los Alamos" above and "Lubbock" below.

The names of tests came in series. "Lubbock" was one of a number of large Texas towns so honored. Jas often mentioned that designers were superstitious about everything to do with their tests, and names could set the wrong tone entirely.

"There was a nautical series, and a failure had the name 'Bilge.' I'm not too happy with one of my upcoming shots. It is in a group of mechanical terms, and mine is called 'Slip.'"

While the long months of scrupulous engineering went on at Los Alamos, crews drilled the hole at the Nevada Test Site. The laser-guided drill carved the sides of the hole straight and clean to amazing tolerances. The geology of the area around the shaft was analyzed to anticipate how the explosion would move through the ground. Before a test went off, scientists tried to calculate within a foot or two the diameter of the crater to come. Trailers that contain recording equipment were set down on platforms just outside the predicted circle, as close to the hole as possible to save money on cable. The platforms' splayed feet rested on an eighteen-inch-high honeycomb of aluminum to absorb ground shock and minimize the pitch and roll during a detonation. "Lubbock" succeeded in smashing the honeycomb down to a couple of inches.

A few weeks before a test, the rack, the diagnostic instruments, and the parts of the nuclear device arrived in Nevada for assembly. All the various sensors were placed in the rack, protected against mutual interference—"crosstalk"—with heavy shielding until the rack was thick with lead and plastic and tungsten. The instruments were hooked to the cables waiting at ground zero. Once this movable laboratory was ready, the explosive was assembled in a bunker. An armored convoy carried the device to the edge of the shaft, where it was inserted in the rack, connected to the electronics that would send the firing signal, and sunk into the hole before nightfall.

According to Jas, "While the device is out in the open, signs come out that say 'The President has authorized lethal force . . .' and the guards you see are not the usual amiable ones, but guys who look like they read the want ads in *Soldier of Fortune*."

The lead designer was in the bunker, watching as the components of the explosive were fitted together. The lead engineer was in charge, but the designer watched for any anomalies. A colleague of Jas's once spotted a material going into the explosive that was the wrong color. "It turned out to be not just the wrong color, it was the wrong stuff—the test would not have worked.

"In the old days, designers would pat a finished assembly—if it felt warm, they knew for certain there was a real plutonium pit inside."

During the previous couple of years, the lead designer ran the show. At this stage, he or she had little left to do but wait. The pros at the test site took over.

Once the rack was painstakingly lowered all the way to the bottom—that was the step in "Lubbock" 's progress I was there to see—the hole was backfilled with alternating layers of fine sand, pea gravel, and sometimes concrete, a process that sometimes took weeks. Finally, epoxy was poured in to fix the cables in place. The elaborate plug prevented radioactive gases from seeping into the air after detonation.

Now the scene shifted from the drill site to the command post, a building on a ridge between Frenchman's Flat and Yucca Flat. The device was ready to go and everyone's eye turned to the weather. The job now at this point was to ensure that any radiation accidentally leaked from the shot would do the least amount of damage. There had not been an accidental emission from a test in the last twelve years, but a shot would be delayed if the wind was blowing toward civilization. Balloons went up to trace the wind, airplanes stood by, ready to track any radiation that moved off-site. Anybody living or working in the area around the site was accounted for in case there was an evacuation. Perimeter patrols looked for intruders and tried to scare wildlife off to a safe distance.

The command center was a small, darkened amphitheater with rising banks of desks curved toward a wall of television monitors. The TVs showed readings from seismographs, radiation detectors, and weather stations, and live pictures of ground zero.

The lead designer was a minor figure here. A woman designer put it this way, "You are like the father at a birth. Someone else is doing all the pushing."

The night before a shot, designers just killed time and worried. They slept in a minimalist cinder-block dormitory in Mercury where some rooms had no air-conditioning. It was often breathtakingly hot in Nevada.

Jas said, "The heat and the dust—you worry about the effects on the equipment, especially the electronics, let alone your own discomfort. But I heard of a Russian weapons scientist saying once, 'If we had such a place for testing, with weather as good as yours, we would beat the pants off you.' They are stuck with arctic conditions."

Since there is nothing left to do, most designers fell back on personal rituals.

Jas: "Designers are the most superstitious people in the world. We don't completely understand the physics of bombs. You never fully know what caused a failure—any one little engineering detail could make the difference. Anyway, if you have had one successful shot, you tend to repeat whatever you did on that occasion. One guy always wears white. One woman was eight months pregnant at the time of her first shot, so she didn't go to Nevada. Now she never goes. One or two might drink a bit more than usual."

A couple of hours before detonation, technicians made one last trip to ground zero to remove safety devices that block the firing signals from the command post. Then a countdown began. Up until the final moment, the sequence could be stopped. A helicopter cruised around the hole sweeping the area with a camera.

"I remember one shot, it was five-four-three- and a flock of crows landed right on top of the site. Of course, they could make a quick getaway. I have heard of the same thing happening, only it was a couple of people, protesters, that the camera spotted running toward the hole. Everything came to a halt while they were arrested. They did not seem to realize they could be badly injured."

There is a reassuring sameness to the tests. The Los Alamos test director gives the final okay to start the firing sequence—a series of remote commands that arm the device and activate the diagnostics. One designer said, "You hear the countdown in real time, then between one and zero, it feels like an hour goes by."

When the thermonuclear device goes off deep underground, it turns nearby rock into a gas, actually a plasma of stripped atoms raging outward in an expanding sphere millions of times denser than its surroundings. After the first microsecond, a shock wave ripples out from the center of this holocaust and races out through the ground, crushing the layers of rock on all sides. Right behind the shock wave comes a balloon of heat, inflating to burn solid stone into a cavern of vapor. The walls of the hollowed-out cavity turn molten, and the gases within remain for a while as thick as the rock used to be, holding the earth in place.

All that one sees of this hellstorm at the surface is a puff of dust and a seismic swell: the ground can buck up eight or ten feet directly above the blast.

As it cools, the liquid rock slides down the curved sides of the cavity into a pool at the base, where it resolidifies into black glass. The pressure of hot gases abates and the roof of the hollow sphere caves in, setting off a rising column of collapse that finally reaches the surface, minutes or hours later, forming a crater. The glass captures weapons debris and the falling chimney of rock blankets the hole and traps the radioactive gases.

During the first nanoseconds after detonation, line-of-sight pipes, metal tubes fixed to the rack and pointed at different portions of the nuclear device, act as avenues for radiation, carrying gamma rays, X rays, and neutrons to the sensors higher on the rack. Detectors may sample the neutron spectrum from different points in the explosion, testing for inconsistency in the dynamics of the nuclear reactions. In the instant before the rack and all the instruments vanish in the fire, the measurements travel through the cables and into the recorders in the nearby trailers.

While this storm rages, the shot's lead designer, standing in the back of the command center, can only watch for the telltale dust cloud and await the first taste of data fed back from the hole. Soon after the detonation, the crew will return to the trailers to monitor the recording equipment. It is impossible to know just when the crater will abruptly sink.

After "Lubbock," the crater spread right to the trailers' feet, jolting the occupants. Referring to the physicist who said there would be no crater at all, somebody yelled out, "She blew it."

In a final culling of information, another team drills down to take samples of the glass from the bottom of the cavity. The radioactive leavings will be returned to Los Alamos to be chemically assayed for fission products. This helps form a picture of what actually happened inside the bomb.

"One of the advantages of underground testing is that it preserves secrets," Jas said. "You can scoop the air after an atmospheric test, even at a great distance, and figure out through back logic what kind of device would produce this array of fission products."

It takes months to collate all the measurements from

a test. What happened may or may not be what was supposed to.

Del said, "Getting surprised in Nevada is what makes our best designers. They learn that their computers can't come that close to modeling everything that happens in the real world."

AFTER 1992, THE NEVADA TEST SITE WAS AS QUIET AS A tomb. An indefinite moratorium on nuclear tests had stopped everything in its tracks.

Testing had always been thought of as an engine of the arms race and opponents wanted to turn the ignition off. Most Los Alamos designers wanted to keep on testing, despite the end of the Cold War. Understanding and controlling the physics of a nuclear weapon was what they did. The way they figured it, you could never know too much or make a bomb too safe. As one weaponeer put it, experiments distinguish reality from opinion. If you are going to handle something as dangerous as nuclear weapons, you had better know how to do it right.

Jas was unhappy with the moratorium. "It will not be easy to be a designer without tests—what scientist wants to work without doing experiments?"

Los Alamos would be charged with the task of maintaining the ability to design weapons without actually doing it. It was like asking an air force to train pilots without using jets.

"The fact is, we still do a remarkably bad job of understanding exactly what is going on inside a bomb. Doing only computer modeling would be a return to Aristotelian assumptions that logic can penetrate all phenomena. Any scientific discipline has to tie itself to reality," Jas said. "When we test, we learn how the computer lies.

There is nothing like practical experience—you learn from doing. Get too many theorists in the room and you don't want them to *touch* anything."

Del was not as worried about the suspension of nuclear testing. "At some point the pursuit of understanding physics loses meaning," he said. "We already know we can make something that will blow up."

RIGHT OUTSIDE THE GATE THAT STOOD BETWEEN THE Nevada Test Site and Highway 95 were two holding pens, barren squares inside high chain-link fencing, about eighty feet on each side, one for men and one for women. These temporary jails were furnished with portable toilets; there was no shade. Each spring hundreds, sometimes thousands, of protesters had gathered nearby to register their disagreement with United States nuclear weapons policy. They were not too happy with the weapons program's poisonous by-products, either.

"They come in the Lenten season and around the anniversary of Hiroshima," Don Collins told me. "The organizers usually sit down with the Department of Energy beforehand and tell them their plans."

They camped across the highway in tents and RVs. A few people would backpack food and water and walk in under cover of night—only miles of bald desert protected the site. Mostly they were there to block the road and cause a little irritation.

Don spoke of the protesters with the same kind of affection he felt for the owls and coyotes. They were part of the scenery for him, a mark of the changing seasons.

"You get a few radicals," Don said. "They make these blocks of wood with nails sticking out everywhere to tear up tires, or they wear studded bracelets under their sleeves that cut the guards when they get arrested. But that is unusual. You see all kinds of people—a lot of retired couples in campers, not crazy, just concerned."

The arrests were ritualized. Demonstrators put their bodies where they should not be, county sheriff personnel handcuffed them with disposable plastic manacles, and bused them into the nearest town for processing. When they were finished, they bused the protesters right back to the campsite to get rid of them.

In the literature that organizers passed out to the crowd were a lot of warnings about radioactivity on the site: hot spots and dust particles ("cover your face"). During the tour, I had seen a few areas surrounded by signs, various versions of "Radioactive Hazard," but not many. Don answered my question about the dangers of radiation with a story. "Not long ago," he said, "I decided to look at the records of my exposure, the readings off my dosimeter since 1956. Now, I have seen atmospheric tests and hot reactor experiments and gone in afterward for cleanup operations, and the records show my radiation age is sixty years. I am fifty-eight, so that means I've picked up two years over the average U.S. background exposure—what a normal person gets from just walking around. I don't think that is so bad."

DON HAD TO TAKE A DETOUR TO RETURN TO MERCURY AT the end of the day. It was slow going over asphalt buckled by heat and, here and there, buried under sandy washes. The desert was tossing the old road off its back.

Karen Randolph spoke up to fill me in on a multimillion-dollar project to monitor contamination in the ground water in and around the site. I noodled a couple of notes but mostly stared out the window. We had been on the move all day, and I still had no sense that desperate things were going on at the Nevada Test Site. The violence was cloaked, the connection to it attenuated by plastic cables and digital readouts. But as Don drove me back to the city at twilight, I kept thinking about pigs roasted in little Eisenhower jackets.

CHAPTER

4

BAKSAN, RUSSIA

> Something deeply hidden had to be behind things.
> —Albert Einstein

VLADIMIR GAVRIN, A RUSSIAN NUCLEAR PHYSICIST, STOOD two miles deep into the heart of Mount Andyrchi in southern Russia, listening to underground streams that surged through seams in the rock only a few meters above the tunnel that housed his laboratory.

"Hear the mountain breathing."

From inside the tunnel the mountain did seem like a living thing. The walls were warm—105 degrees F—heated by the earth's primordial furnace. The pressure of the water flowing through the hidden aquifers constantly threatened to breach the metal walls of the laboratory.

Gavrin was a vigorous man with a pugnacious jaw and deep-set eyes. He liked to interrupt himself in the middle of explaining some bit of science to recite strings of Russian verse, going on and on with rolling R's and sonorous vowels until he reached the end of his memory. But he didn't look like a poet—he looked like a boxer, with combat lines etched deep in his face.

Gavrin's lab was the site of SAGE, the Soviet-American Gallium Experiment, manned by an international group of chemists and physicists, including a team from Los Alamos. It was the first time an American weapons laboratory had collaborated on a Russian experiment. Los Alamos signed on to the partnership in 1986, when Gorbachev was just coming into power. The icy confrontation that was the Cold War had yet to thaw, but the Soviets had already invested so much money in the work, politics took a backseat to expediency.

The Russians had already taken the trouble to dig a cavern out of a granite mountain and to stockpile a costly rare metal for the heart of the experiment. The Americans pitched in with computers, fiber optics, and various high-tech sensors. SAGE was essentially an underground telescope designed to look at a puzzling subatomic particle, the neutrino. Experiments in subatomic physics commonly depend on accelerators, enormous machines that push particles along at unimaginable speeds and smash them into targets. But SAGE was a little less orthodox: the scientists stood still and looked at particles the sun delivered in abundance and for free. It was astrophysics in a hole in the ground.

"There have been twenty years of battles to finish this laboratory, many tests of new methods—some setbacks, some successes—some people gave up and stopped believing in the work. I pushed maybe too hard sometimes to finish this laboratory," Gavrin said. "Sometimes we were pouring concrete for the floor and testing the chemical apparatus at the same time. The theoreticians said I was too impatient. Somehow, I usually got done what was necessary, like a small animal who becomes very fierce in his own territory and frightens animals twice as big."

There was no expectation of a practical payback. The experiment was fundamental science, the kind of science that asks not how to make the world better, but poses the question: What is the world made of? Its very impracticality is no doubt another good reason why the collaboration did not scare off either of the two governments.

For Los Alamos, it was typical of the projects that sprouted up here and there. Dollars spilling over from the weapons budget fell into the hands of smart, industrious people who had all kinds of ideas about what to do with them. At bottom, bomb work was understanding the subatomic world, and once scientists stepped inside that world, there were endless places to go exploring.

The two superpowers competed both on and off the playing field of deterrence. American triumphs in basic science were weapons of intimidation, a kind of bluster keeping the Soviets in fear of our superior powers of intellect. This was part of the appeal of funding science, along with the distant promise of some useful payoff.

I FIRST ENCOUNTERED GAVRIN IN NEW MEXICO WHILE he was attending a meeting of the SAGE scientific team. The group gathered at a hotel in Santa Fe. To set foot on laboratory grounds, all foreign nationals, but most especially Russians, needed escorts, dispensations, affidavits, approvals from Washington, forms in triplicate—a general bureaucratic wash-and-tumble-dry. On this occasion, there were too many Russians and too little time, so they met off campus.

Most of the Russians were young men in Levi's and Reeboks who were almost indistinguishable from the Americans in the room. They went to the shopping mall on their breaks.

I wanted to go to Russia to see SAGE, and I would need Gavrin's intervention to make it possible. While his colleagues wandered around in Montgomery Ward, he stayed behind to work through lunch. A Los Alamos team member guided me up to Gavrin for the introduction. Could a visit to the experiment be arranged?

"A journalist?" He looked at me sternly. "You can come stay with us as long as you cook."

Gavrin never missed a chance to take a swipe at Western ethos, in this case, feminism, and he was no kinder to Soviet habits of mind. But even with a face like thunder, he said these things to get a laugh.

Gavrin has used his regular visits to the United States and Western Europe to experience everything he could of foreign culture. He has drunk whiskey in a strip joint on Times Square and bought forbidden books—Solzhenitsyn —in Switzerland.

"It was a rule several years ago that when traveling abroad it was forbidden to go to church. You had to sign a paper before you left Russia," he told me some time later. "That made me curious. I heard a sermon on world peace in England. In Italy, I followed a line and took communion by accident. In Wiesbaden, after midnight, I saw through the door and went into a church to see one old woman kneeling. I stayed in the back and did not make a sound and for once I felt a stirring in my heart."

MOUNT ANDYRCHI ROSE LIKE A DOGTOOTH OUT OF THE narrow valley of the Baksan River in the Caucasus Mountains, where Europe shaded into Asia. It was a sharp volcanic range that formed a bridge of land between the Black Sea and the Caspian, a small appendix to the vast Russian Republic.

Baksan Valley and the apron of flatlands at its base made up Kabardin-Balkaraya, a country of dark-eyed horsemen and herders. They were Muslims, a part of Russia without being Russian.

The Caucasus Mountains seemed restless on the surface—the sound of rock falls often rumbled in the night—but deeper down, they were geologically stable. When Gavrin and his mentors at the Institute for Nuclear Research in Moscow were seized with the vision of underground physics, they brought their ambitions here. Where there was nothing but a barren stretch of road, the scientists drilled into the mountain on one side of the valley and across the Baksan River built Neutrino Village.

"Certain officials could not believe that we would excavate such a huge tunnel just for science," Gavrin told me. "One time the miners were too lazy to fix some lights in a small passageway, and the visitors were certain that was where the secrets were buried."

The village itself was a town of concrete block offices and apartment buildings tacked on to a sheer, treeless slope. The miners, electricians, carpenters, secretaries, and day laborers whose work supported the tunnel and its three laboratories, including SAGE's, lived there year round. They numbered close to one thousand, including some four hundred children. Most of the SAGE scientists lived in Moscow, but they came to Neutrino Village for two weeks of every month to perform the chemistry that tracked down solar neutrinos.

The journey to Neutrino Village started with a

crowded, wide-body jet into Mineral Water (Mineral'nye Vody). The airport was bedlam. Hundreds of passengers, backed up from days of delayed and canceled flights, hunched over their belongings in the cavernous cold of the waiting room, or, using their baggage as a battering ram, shoved toward gateways and ticket booths.

From the airport at Mineral Water to Baksan, ground transportation was either a five-hour bus ride or a taxi. Gavrin cadged gas coupons for months to ensure American visitors a private car. His job as director of the laboratory went far beyond physics. Each month he organized two weeks of food, airplane tickets, work schedules, tools and materials for his research team.

Yura, my driver, sped out of the airport under a gray drizzle. He flicked on the wipers just a few strokes at a time, as if to spare the badly cracked windshield any undue strain. After a quarter of an hour of driving, he made a left off the main highway to stop at a park filled with schoolchildren and old men in fedoras. Yura turned to address me with various words and gestures, inviting me to get out of the car. He spoke no English, and the Russian in my phrase book did not help to sort out the situation. I gave up and followed him through the rain to the base of a statue. It was a monument to the poet Lermontov, a native son.

"Lermontov," Yura said.

"*Spa-see-bah*," I replied. Thank you.

Back on the narrow road to Neutrino Village, we zigzagged upward toward Mount Elbrus, Europe's unsung highest peak. The road was more a ledge than it was solid ground, with a hundred-foot cliff defining the outer lane, an arrangement that might have induced caution in a lesser driver. But Yura had the reflexes of a fighter pilot. We swerved around broken trucks, skidded across washed-out pavement, and plowed through a herd of sheep that poured off the hillside, all without wasting precious resources like the brake pads.

We reached the village in the late afternoon, sliding to a halt on a thin layer of snow in front of the scientists' cottage. I got out of the taxi and noticed a black bull above me on the hill—he was mouthing the latch on a Dumpster, patiently trying to open the lid and free the contents. I heard the Baksan River's white water crashing against the rocks below. Through a heavy mist, the underbelly of a winter cloud, I could see a footbridge over the river and, on the far side, huge iron doors to the tunnel that housed one of the most sophisticated particle physics experiments in the world.

THERE ARE MORE NEUTRINOS IN THE UNIVERSE THAN ANY other form of matter. They are everywhere: torrents of neutrinos flood through you every minute; neutrinos fill every corner of the universe, yet almost nothing was known about them.

The neutrino is so vanishingly small it was considered massless—with no heft, no weight, just a packet of energy so elusive it was like movement glimpsed from the corner of the eye. When you looked straight at it, it did not appear to be there at all. SAGE was finding tantalizing evidence that neutrinos were more than mere phantoms. They seemed to have mass. A neutrino with mass: if it were true, it might prove to be the particle that weaves the fabric of the universe, describes its beginnings, and determines its ultimate fate.

SAGE's Mount Andyrchi and the sun that shone on it were each vital parts of the machinery of the experiment.

The sun acted as a powerful thermonuclear reactor, creating billions of energetic neutrinos every second, as well as light and heat, as it burned. The mountain was a thick shield that absorbed almost all radiation except the ghost wind of neutrinos. Neutrinos glided through it as if nothing were there.

These solar neutrinos were extremely hard to catch, but because there are so many of them, there was an infinitesimal chance that one would bump into something. When that happened, the neutrino performed alchemy, changing one element into another. A neutrino would merge with a neutron in the nucleus of an atom and transform the neutron into a proton, totally altering the chemical identity of the atom.

The "telescope" at SAGE was sixty tons of a strange metal, gallium, stored inside eight huge vats. The nucleus of an atom of gallium 71 has 40 neutrons; 60 tons of gallium contain 10^{30} atoms (one with thirty zeros after it)—a big target for neutrinos. When a neutron inside a gallium atom engaged a passing neutrino and turned into a proton, the new atom was radioactive germanium, an element that could be chemically separated from the surrounding gallium and counted, atom by atom. When I went to Neutrino Village, SAGE had been looking for germanium atoms—evidence of neutrinos from the sun—for about three years. In that time, fewer than half the neutrinos the sun should supply had been found.

Let's say you walk into a bare room and find a set of children's blocks scattered on the floor and in the corner a box marked "Blocks." You pass the time by stowing the blocks in the container provided, but when you get to the end of the task, you see that the box is not filled, that there is an odd gap between the blocks and the top of the box.

Now you know something significant: either you have the wrong box, or there are more blocks somewhere else.

In the neutrino experiment, the size of the "box" was determined by what is known about solar activity. The sun has been measured many different ways, all of which agree pretty closely on how many neutrinos its reactor core should be producing. Because the box looked right, experimenters began looking for missing neutrinos to fill it to the top.

What happened to all those neutrinos? Current science cannot account for the gap, but there are theories that go beyond the bounds of proven physics. The most elegant explanation of the missing neutrinos predicts that, in their journey through the sun, many will "oscillate," that is, transmute into other, more exotic, species of neutrinos. The gallium at SAGE would not react to these new particles, so the missing neutrinos could be there—but hidden from view, requiring new ways of seeing to find them.

The oscillation theory calls for a revolution in physics. All the mathematics that currently describes the universe assumes that neutrinos are massless. If neutrinos are oscillating, they must have mass. So neutrinos, among the most common stuff of the universe, seem to have a robust personality that scientists believe could illuminate the deepest cosmic mysteries.

A TRAIN RAN THROUGH THE DARK AND SULPHUROUS tunnel, iron wheels screaming along the rails. At the end of the line was SAGE, slick and clean as a new-minted coin. A plump stoic, the classic Russian babushka, guarded the door, making sure everyone changed from outdoor to indoor shoes before entering.

Inside, steel vessels, like giant Cuisinarts painted red, green, and yellow, squatted on the floor of the lab. They held the gallium, warmed to a sluggish, gray liquid. Inside the tanks, motorized paddles stirred the metal at a constant speed, adjusting to any changes in viscosity. At room temperature, gallium looks and feels like the graphite in a No. 1 pencil, but it melts in the palm of the hand. On the occasions when bricks of gallium were transferred to the vessels, the workers' skin and clothing were thoroughly blackened by the end of the day.

Sixty tons of gallium: the bricks would stack up to the size of a one-car garage. Each month the scientists processed the metal chemically to extract about twenty atoms of germanium from the 10^{30} atoms of gallium. It is like searching for a few specific grains of sand on the beach at Waikiki.

The apparatus for the extraction looked like a tabletop chemistry set blown up to gargantuan size. Test tubes six feet tall connected by great elbows of glass hung from a high steel scaffolding. Gavrin called it "our chemical jungle." The main room was all business, but in an alcove to the side, where a delicate maze of blown-glass tubing handled the later stages of processing, someone had taped a Beatles poster to the wall.

There is an art to drawing the few germanium atoms out into the open. Chemists added what Gavrin described as "aggressive reagents," solutions of hydrochloric acid and hydrogen peroxide, to the vats. The reagents bonded with any germanium and, holding it, rose like cream to the top of the gallium. The timing, the temperature, the proportion of reagents, the speed of the paddles that stirred the metal—all the parts of the recipe have evolved through years of testing, some not so successful.

"At the beginning, we worked in our laboratory near Moscow," Gavrin said. "We put in the chemicals and I stood right there and watched them work. At that time, nothing was automatic. I would wait until the gallium looked right and then say, 'Stop.' One day, we had a new vessel with a window three centimeters thick bolted into a stainless steel ring. The first time we tried to do the process with the vat hermetically sealed, I was peering in the window and heard a sound like a shot. A blast knocked me over, the steel ring hit the ceiling two-stories up, glass flew everywhere. We made a lot of guesses but never really figured out the cause of the explosion. Nevertheless, the vats here have ventilation."

The Baksan laboratory was about two hundred feet long, thirty feet wide, and forty feet tall. Behind the high, vaulted space that held the vats, there were several smaller rooms and a staircase to a second half-story. Igor Knyshenko was one of the chemists who monitored the vats and dials at far ends of the lab during the extraction. He dashed up the stairs to a loft where a bank of computers registered data; he descended at a run, stopping for a moment to peer through a hatch into the molten gallium; he raced over to tap on the tubes where cascades of liquid washed down the inside of the glass. The procedure took twenty-four hours from beginning to end. The young Russian chemists spelled each other through the night. For at least half those hours, Gavrin was at a desk in the middle of it all, troubleshooting, tracking every nuance of the chemistry.

Inside SAGE, the light never changed, the air was still—twelve hours could pass without notice. More than a mile of mountain separated the sky and the cambered roof of the laboratory, keeping out a barrage of natural radiation from cosmic rays. Cosmic rays are like a gale from deep space that strike the top of our atmosphere, raining down penetrating subatomic particles. The atoms in all that dirt above the laboratory absorbed most of this unruly mob that would otherwise disrupt the gallium.

The gallium had to be clean and quiet at all costs. Only neutrinos could be allowed in to disturb it. Stray radiation also turned gallium into germanium and caused false readings that could overwhelm the neutrino numbers. Because another source of trouble was radioactive impurities in the mountain itself, the laboratory, a metal shell fitted into a cavern, was surrounded by one last radiation roadblock, a layer of concrete that filled the space between the walls and the jagged arch of granite above. The concrete sealed the room off from traces of uranium in the rock.

On a late November evening, Gavrin sat at the long dining table in the scientists' cozy two-story cottage in Neutrino Village and talked about building the lab. We drank strong tea sweetened with apricot jam. Gavrin had tried but failed to find cheese and cooking oil that week, and one day there was no bread, but the meals were rich and unhurried anyway. Natasha, the cottage's general factotum, could cook a rabbit ten different ways.

"The concrete itself could not be radioactive," Gavrin said.

He often paced when he talked, but that night he was tired after spending all day underground. Fortunately, he was never too tired to tell a story. Gavrin sat with his knees almost touching mine, leaning forward, as if to be sure that I was listening to every word.

"We investigated many places looking for the right rock and sand. We chose dolomite in the Urals, very old rock. And from the Ukraine very pure quartz sand.

"But to crush the dolomite, we needed a machine that was completely clean. I found a factory that could spare time on one of their crushers. We cleaned it very thoroughly, and then I posted a man by the machine to see that they did not put any other rock but ours into it.

"Even carrying it into the tunnel, it was necessary to have cleaned carts that would not contaminate the mixture. Many things we carried by hand. We would even wash down the walls of the tunnel—all three and a half kilometers—before we transported our materials. The miners thought we were very strange."

In the late 1970s, when the Russian physicists decided to use gallium's abundant and cooperative neutrons for detecting solar neutrinos, there was not enough of the metal in the whole world to satisfy the demands of the experiment's design. Estimates of the cost of purifying sixty tons of gallium were in the range of $30–$40 million. But scientists argued that the cost was justified: the neutrons in gallium were one hundred times more likely to capture neutrinos than the neutrons in less expensive elements.

It was Gavrin's job to persuade his government to spend the money.

As we talked past midnight, many years after the events, Gavrin's feelings were still fresh—his resigned sadness about human nature, his pleasure in his own cunning.

"About something of this importance only the very top of the Communist Party, the Politburo, could make a decision. It was the Minister of Colored Metals, in charge of the production of such things as gold and copper, who could order the production of gallium. It was comparable to starting up uranium production, but that had been easy because Stalin had wanted it. Only the physicists wanted gallium.

"Gallium is produced when you refine aluminum. We

searched all over the Soviet Union and found a factory where the technique for gallium was very advanced, but they were only making a few kilograms. We had many meetings with technologists and scientists and deputy ministers to campaign for everyone's support before we went directly to the Minister. Everyone was enthusiastic but I saw two difficulties. At that time, people would advance their careers and earn better money if they fulfilled the Party's plan. Doing something new, making improvements, did not necessarily bring benefits.

"The second problem was that we were asking to produce just a certain amount of gallium for one purpose and then there would be no future. So I searched for other applications for gallium. One day, I read in a Western newspaper about the capture of Che Guevara. The story was about how American spy planes used infrared optics to see Guevara's campfire in the Andes. Infrared optics, I knew, need gallium.

"The next meeting with the ministry bureaucrats, there were military men invited to listen to this information about Americans using gallium. They did not introduce themselves; I never knew who they were. I just know that after that meeting, interest in gallium was much greater."

I found it difficult to associate this methodical diplomacy with Gavrin. He was a man who crackled with energy. His anger and his wit could snap out at any moment. When he was surrounded by his coworkers, I noticed they usually kept their eyes on him, as one would on anything that might be a little dangerous. Apparently, his passion for gallium astrophysics had invested him with the necessary patience.

"Then it was finally time to make a presentation to the Minister of Colored Metals. I went to the meeting with my chief, Zatsepin, and Markov, a very prominent scientist in the Soviet Academy. As we sat before him, I could see on the Minister's face the struggle to make a decision. Would it bring him more power to reject the important scientist or to take the job of gallium production with many possible problems? After all the preparation, it took only a moment—he said yes."

In the early 1980s, the half-dozen American scientists on the SAGE team had been working through their own bureaucracy, trying to get their hands on enough gallium, but they failed where Gavrin succeeded.

I HEARD A STORY, PROBABLY TRUE, ABOUT A LOS ALAMOS scientist who was reprimanded for scribbling "Eat Neutrons, Ivan!" on a nuclear test device that was on its way downhole in Nevada.

Los Alamos weaponeers fixated on the Russians. Bookstores in town were packed with volumes on Soviet history and current events. For the Americans, Russian weapons scientists were archrivals, respected equals, mortal enemies, yet members of the same small fraternity: the artisans of mass destruction. They were also virtually inaccessible. American and Soviet weaponeers might brush up against each other in the course of arms control work, but for the most part, they were a mystery to each other.

There were no doubt scores of weaponeers who longed for a glimpse of the Evil Empire, but when the Department of Energy relaxed its rules for the first time and permitted a group from Los Alamos full-blown commerce with the Soviet Union, it was a group from one of the sideline sciences, particle physics.

Tom Bowles, a neatly dressed, bespectacled physicist from Los Alamos, was the leader of the American team. He was also at SAGE that November. While we talked in one of the cubicles off the main laboratory, he used a pair of needle-nosed pliers to tinker with the torpedo-shaped box of electronics that makes the final count of germanium.

"Sometimes the equipment we brought over was on the embargo list and we had to get special licenses to let it go to the Soviet Union. The Department of Energy has audited our records of the collaboration numerous times over the years. Even though they authorized it, the bureaucrats look at all the money and equipment we send over here and they ask what's going on."

The device Tom was repairing, a quartz counter, weighed a thousand pounds and hung from a ceiling crane. He had to get down on his knees to reach the part he was fixing. He was as well acquainted with the mechanics of the laboratory as he was with the physics of neutrinos, and, like Gavrin, Bowles never sat still.

"It is not the same as working with an experiment in, say, Western Europe. When we first came, there was only a fifteen-year-old computer at the lab," Bowles recalled. "It took days to get phone calls through from Moscow. When you are here, you can't just go to the store to get the right-sized nuts and bolts. You have to plan your moves months in advance.

"It took a while, but we are getting used to the differences between the two styles of science and making the most of them. Russians have long-term planning, steady resources. Americans are more opportunistic. We react more quickly to contingencies. Russians have a step-by-step approach.

"We had argued that the Russians should move the equipment from the test lab in Moscow to Baksan sooner than the plan called for. We Americans wanted to get started, to get the equipment running, find the problems, and solve them. They finally went along with us. And, of course, we relied on their perseverance, their ability to acquire the gallium and maintain the lab.

"Neither group could have carried this out so well on their own. We have made the most of each other's strengths."

For the first couple of years of the collaboration, there was always one American or another at the site. Now, Bowles and his associates were making the two-day trek from New Mexico to Baksan three times a year to work on upgrades. The scientists were tweaking the system, trying to get the data closer and closer to certainty. They upgraded the hardware and sharpened up the software, trying to identify and count only the germanium created by solar neutrinos, and to disregard any germanium atoms that might arise from other causes.

Distinguishing the meaningful signal—the neutrino-made germanium—from all the noise created by background radiation was a process they perpetually refined but never perfected. The task was something like standing by the railroad tracks, listening for a violinist playing in the dining car as the train roars by at eighty miles per hour. It took lots of equipment and effort to filter out all the sounds but the Mozart.

"We are turning this into a clean room and a Faraday room for the quartz counter that registers the germanium," Tom said, tapping the intricate package of wires and semiconductors.

"That means the air will be filtered and pressurized to keep out dust; and it will be electrostatically protected. When it is finished you will pass through a foyer and put on protective clothing before you come in."

Tom Bowles is careful with his enthusiasms. He was smiling but almost solemn when he told me, "I never thought I would be this lucky. You always expect to get ambiguous answers. I was shocked as hell that the results were so clear cut. The exciting part is that the experiment sheds light on the very smallest and the very largest phenomena in the universe."

IT IS AN ARTICLE OF FAITH IN PHYSICS THAT THE WORLD'S bewildering mask of complexity hides an ultimate simplicity. A Greek philosopher, Democritus, first proposed the idea that the world was made of very small, identical, irreducible bits of matter: atoms, from the the Greek word for unbreakable.

Scientists have searched for this singular grain of matter, but nature has not cooperated. Atoms are not smooth and indivisible. They have two kinds of working parts—quarks and leptons. Quarks are the units that combine to make protons and neutrons. The lepton family includes electrons and neutrinos. And inside the tiny confines of the atom, there are no fewer than four separate forces in action.

Atoms seem as complicated as a game of chess, their pieces odd-sized and moving across the board with different, staggered gaits. Yet even when confronted with this proliferation of nuclear particles and nuclear forces, physicists refused to abandon the ideal of simplicity.

Theorists choose to see all the mismatched parts of the subatomic world as the shattered, cold remains of an earlier unity. They look back to the time of the Big Bang, when the universe was so furiously hot that all the particles and forces melded into one reality where they were not distinct but identical with one another.

The idea of temperature transmuting matter has an everyday counterpart. With enough heat, water appears only as steam. After a bit of cooling, steam condenses into liquid, which has properties completely different from steam. Get it cold enough, and water turns to ice, which seems to have no connection to the the original steam and is not at all like liquid.

Most physicists believe that the universe was simple at its superheated birth, but fractured into multiple parts and forces as it cooled. Grand Unified Theories, or GUTs, are mathematical pictures of how three nuclear forces ought to blend into a single force at transcendentally high temperatures, a force that affects all particles uniformly, erasing their differences. The elementary bit of matter Democritus postulated is an artifact of the most distant past. Grand Unified Theories are like artists' renderings of the creation of the universe.

But a theory without laboratory evidence is just a guess. Theories need to be fed facts.

GUTs require that at some deep level, quarks and leptons must share an identity that connects them to their common past. This kinship is possible if neutrinos have mass. Data from SAGE provides the first solid clue that can verify a Grand Unified Theory. Without SAGE's results, GUTs are acts of informed imagination. Afterward, they are measures of reality.

Theories are also open to differences of interpretation. When I mentioned Grand Unified Theories and the beginning of time to Gavrin, he said, "You can discuss this topic only with God."

A neutrino with mass hints at profound truths about the quantum world of the atom and about the Big Bang, but that is not the end of its possibilities. This newfound neutrino might help measure the cosmos. Astronomers understand that the great mass of the universe, 90 percent of it, is completely invisible to us. We actually do not know what most of the universe is made of—the stars are just the raisins in the cake.

By observing the movement of radiant objects in space, scientists deduce that the seemingly empty stretches of the universe are actually full of matter. The clockwork whirling of the galaxies cannot be explained unless the stars we see are embraced by a cloud of matter we cannot see. Suppose you watched a movie of a brick falling off a wall and slowly drifting to the ground. You would probably conclude that the brick had dropped through some heavy, transparent medium, like water. Watched closely, stars move as if they are swimming in an unseen ocean.

The invisible stuff is called dark matter. Nobody knows what dark matter is, but a neutrino with mass is an attractive candidate to explain at least some of this mystery. Even a neutrino with a tiny mass would add up to a formidable amount, since there are more than six hundred neutrinos for every cubic centimeter of the universe.

When we understand dark matter, we can begin to comprehend how stars and galaxies pulled themselves together and resisted the relentless force of the explosive outward expansion of the universe as a whole.

Dark matter is more than a small technical problem associated with the birth of galaxies. When we figure out dark matter, we will know exactly how much mass there is in the universe. With that number in hand, we can predict which way the universe will come to an end. While there is no comfort in these prophecies, there is grandeur.

At or below a certain critical mass, the universe will continue to expand for eternity, growing colder, thinning out to a black, meaningless vapor. Above that critical mass, the universe will expand only to a point, and then gravity will pull it inexorably back together. Galaxies will collide. Stars will collapse into black holes. The universe will speed inward, in a hot, tumultuous implosion, a reverse of the Big Bang, finally reaching a point, a singularity where the laws of physics collapse. The neutrino may hold the power of cosmic annihilation.

THE AEROFLOT CLERK AT MINERAL WATER AIRPORT, A pretty brunette with bright red lipstick, told Gavrin that our return flight to Moscow was going to be twenty-four hours late because there was no fuel.

The Intourist lounge, furnished with sprung vinyl chairs, was badly lit and barely heated. Our group—Gavrin, half a dozen Russian scientists, Tom Bowles, and I—settled in to wait. The room was crowded, but as we watched there were small shifts in the population. Two East Germans in scratchy wool suits left and were replaced by Asians. One or two Africans drifted through as honeymooning Muscovites held hands and strolled. Most of us, though, were there for the long haul and even the Americans absorbed the ambient Russian patience and grew still.

Hours passed and we sank deeper into our heavy coats. Tanya Knudel, a SAGE chemist, gave me a recipe for napoleon pastries. Bowles and Gavrin murmured over a sheet of numbers. Stinging tobacco smoke hung like a second, viscous ceiling, just below the fluorescent lights. The snack bar opened for a ten-minute stint and the room erupted. We had some time ago eaten the sandwiches brought with us from Neutrino. We were hungry, but supply and demand at the snack bar were hopelessly mismatched. Tanya and Ilya, a wry chemist who wrote poetry in his free time, left the lounge with a handful of rubles. They returned in an hour with a sack of apples and another of raisins.

Gavrin from time to time made a visit to the Aeroflot clerk, a woman he had dealt with many times before. He would pull small packages out of his briefcase—I guessed perfume and vodka—and carry them half concealed to where she stood at the counter. By nightfall, he persuaded her that at least the American contingent should board the next flight—a plane that had materialized sooner than expected. Suddenly, it was time to go. In the dark, skating over black ice, Tom Bowles and I rushed across the tarmac to beat the jet's unexpected departure.

Aboard at last and loosened by relief, Tom and I chattered away, enclosed in our foreignness. Tom looks somewhat soft and bland. His mustache is thin and his hair is parted in the middle, giving him a prim, Edwardian aspect. His appearance conceals a knockabout character. He practices karate and spends his vacations being a cowboy on a friend's ranch. He bought his house unfinished and nailed it together by himself when he got home from work.

Tom is from Colorado, the true West. "I grew up watching Dad working with his hands. I also watched too many science fiction movies—by the time I was sixteen, I knew I wanted to be a nuclear physicist."

He earned his doctorate at Princeton. "You had to pay twenty-five dollars for the paperwork to get an official Master's degree, so I just went straight to Ph.D."

Princeton's cyclotron had the reputation of operating in spit-and-baling-wire style. "I studied under my hero, Gerry Garvey," Tom said. "We had to know how to repair the cyclotron, how to design and build our own equipment, and how to do all our own computer programming. I went on to work at Argonne National Laboratory doing, among other things, sodium iodide detectors, which, by the way are also used for nuclear bomb diagnostics."

The stewardess offered us a choice between a half glass of sparkling water and a half glass of tap water. Aeroflot's seats, upholstery, carpets, and overhead storage bins were familiar. Except for the entryway, which passed through a cargo hold and up a staircase to the passenger cabin, it was like every jet I had ever been on—except for that and except for the extraordinary amount of rattling. Over the noise, Tom continued with his personal history.

"The physics division at Los Alamos decided it wanted a program in weak interactions and, in nineteen eighty, hired me to start it," he said. "I spent the first six months thinking, then decided to tackle neutrino mass."

When protons and neutrons switch identities, that is a weak interaction. The neutrino is always a player in the process.

"Los Alamos has a history of balancing basic research with applied science. From the beginning, Oppenheimer wanted the lab to be like a university—satisfying intellectual curiosity at the same time it kept a base of science that the weapons program could draw on."

Being at Los Alamos gave Tom access to tritium, a radioactive weapons fuel that he put to use in an earlier neutrino experiment. The SAGE team was now in a position to give something back to weapons work.

"At SAGE, we are developing high-speed data acquisition and pattern recognition. The IT division [responsible for technology applied to international control of nuclear weapons] thinks it can use the techniques."

Finally, Tom and I were too tired to talk. It was nearly

midnight when we landed in Moscow. We bought sandwiches from a vendor who stood outside the airport grilling meat in the freezing night.

I SAW GAVRIN NEXT IN NEW MEXICO. HE WAS WITH A small crew from Moscow drinking Mexican beer and eating carry-out fried chicken in a hotel room in Los Alamos. They had all slipped their shoes off, as Russians do at home, and they were watching *Star Trek* on television. Tanya Knudel, the only woman on the science team, was making tea in the kitchenette.

The Soviet Union no longer existed and the Russian economy was coming unglued.

"We have to take great care with our supplies. If anything is lost, it cannot be replaced," Gavrin said. "Every month we need two tons of chemicals for the extraction. The reagents are brought in by trucks that need gasoline. It is more than a hundred thousand rubles each time."

Tom Bowles said later, "Gavrin used to call to some Ministry and ask for tools and chemicals, and now the Ministry is lost in confusion. Even if he can find the bureaucrat who used to order the materials, the guy never knew himself where the goods were coming from. There is no natural system of distribution to replace the centralized methods that existed. No one knows quite what is allowed, so suppliers do nothing rather than risk error."

The Russians were in the United States to hash over recent results from Baksan and put the data into final form.

While Gavrin was in Washington, D.C., he discovered that Americans were not optimistic about their economy either.

"I met with an official of the National Science Foundation, to give him information about a grant application that would benefit SAGE. He took two hours to tell me his own difficulties with money. Like the Russian saying, I went to buy wool and came back with no hair."

MORE THAN A YEAR AFTER MY TRIP TO THE BAKSAN Valley, Tom sat tilted back in a desk chair in his Los Alamos office. "The SAGE results are now in a gray area. The neutrino deficit is significant but not so clear that the new physics is definitive. The results are consistent with new physics, and neutrino mass is highly likely, but what does 'highly likely' mean—ninety-five percent or ninety-nine percent? Physicists—you have to move heaven and earth to convince them of something new. And here, the implications are so important, so fundamental, making claims causes a lot of skepticism."

Tom was working on the definitive calibration of SAGE, a procedure that meant creating an artificial source of neutrinos. With a known number of neutrinos, the apparatus and the various mathematical techniques that produced final figures could be cross-checked. The neutrinos were going to come off a highly radioactive lump of chromium 51. The chromium would be manufactured in a breeder reactor in Kazakhstan. "It used to produce plutonium for nuclear weapons," Tom said. "The guys at the reactor are very expert and, at the moment, they have a lot of time on their hands."

The fall of the Soviet Union had complicated matters. Tom said, "Kazakhstan now has its own currency and the chromium will have to cross a border that did not exist when the project began. I am going to have to rely on Gavrin to make it all come out right."

Tom, in pursuit of the truth about neutrinos, was also

deep into another prodigious underground experiment, the Sudbury Neutrino Observatory or SNO. When it is completed in 1996, SNO is going to consist of a clear acrylic vessel, an enormous bottle, filled with a thousand tons of heavy water (D_2O). Heavy water captures all species of neutrinos. The experiment is under construction in a mine shaft in Canada. SNO will be able to distinguish among types of neutrinos and determine where they came from, whether it is the sun or somewhere more distant. The results will finally answer the question of neutrino mass.

"There are different approaches to high-energy physics," Tom said. "One is to build a huge instrument to create the energies where these particles exist. Or you can use high-precision measurements of small phenomena that reflect physics going on at even higher energies. Ultimately, the second approach is necessary because the most exciting physics are at such high energies, they are not attainable with machinery.

"High-precision experiments like SAGE and SNO require a huge amount of data—many events of neutrino capture in very large detectors. You need very sophisticated techniques that deal with instrument unreliability and, always, you have to beat the background down."

At Sudbury, the vessel will be surrounded by specially developed concrete two hundred times lower in natural radioactivity than ordinary concrete. The underground experiment is going to be the least radioactive place on the planet. At the Nevada Test Site, giant earthworks were built to contain man-made radiation. In Baksan and again in Sudbury, the earth is a blockade to keep out the constant shower of radioactive cosmic junk.

"SAGE has begun to zero in on a new theory," Tom said. "SNO will be the smoking gun."

CHAPTER

5

VA HOSPITAL, ALBUQUERQUE

> About all we know about consciousness
> is that it has something to do
> with the head, rather than the foot.
>
> —Nick Herbert, *Quantum Reality*

ED FLYNN AND HIS WIFE WERE IN THE CAR ON THE ROAD between Los Alamos and Santa Fe when her heart stopped. The highway cuts across empty, rolling hills speckled with juniper trees toughing it out in the hard desert sun. It's Indian land, the Pueblos of Pojoaque, Tesuque, and San Ildefonso. The ambulance that finally arrived came from a dozen miles north and carried no special equipment for resuscitation. Only Ed's efforts at CPR kept Gail Flynn alive.

Ed's wife had been taking diuretics, part of a plan to control her weight. As a result, she lost so much potassium her body simply shut down. In the long, devastating moments after her heart failed, so little oxygen reached her brain that it too gave out. Mrs. Flynn receded into a coma that lasted until her death from pneumonia a year and a half later.

As Ed sat at her bedside month after month, his grief was complicated by the mystery of her affliction. Ed was a nuclear physicist at the lab—his whole working life had been spent there, probing the arcane behavior of protons and neutrons. He was accustomed to understanding the way things worked to a high degree of precision. Now, no one could tell him what was wrong inside Gail's head and no one knew how to repair her. He discovered during the final, silent days of his wife's life that the science of the human brain was still sketchy and unformed.

After Gail died, Ed used the proceeds from a prize in nuclear physics to retreat to the Hahn-Meitner Institut in Berlin. He spent a year in the library of the Free University, pondering contemporary research in neuroscience. He wanted to find a way to muster both his physics and the broad, flexible resources of Los Alamos for the task of reconnoitering the brain. His attention finally settled on SQUIDs, Superconducting Quantum Interference Devices. Ultrasensitive superconducting sensors based on SQUIDs are capable of measuring the faintest magnetic fields.

The human brain is a mass of neurons, at least 100 billion individual nerve cells. That is about the number of stars in the Milky Way Galaxy. It is the most complex structure in the known universe, not simply because of the number of cells, but because of the way they interact to create sensation, emotion, imagination, movement, attention, memory, and the ineluctable quality of self-awareness. Each neuron can have thousands of synaptic relationships with other neurons. Each neuron has a range of possible states of activation. The total number of possible neural

VA HOSPITAL, ALBUQUERQUE

states is a number incomprehensibly larger than the number of elementary particles in the entire universe.

Brain activity is a combination of chemical releases and electrical responses that behave like an automated orchestra. Nerve cells of many types, like reeds, brass, and percussion sections, make up the orchestra. As a single instrument sings its distinctive tone, neighboring neurons heed the sound. They either trill a measured note or fall silent in response, and thus the music of nerve impulses is handed off from the woodwinds to the strings. Without a conductor, without a score, the ensemble creates a coherent symphony of mind.

Nerve impulses moving down the length of the neuron—like an electric current in a wire—generate a magnetic field. When a group of active neurons is synchronized, the neurons produce magnetic fields strong enough to be detected through the skull. Still, the brain's magnetic fields are incredibly tiny, a billion times smaller than the earth's. It takes the exquisite sensitivity of SQUID technology to measure them.

Geologists were using SQUIDs to decipher subterranean rock formations and the Navy used them for submarine detection. A few researchers had applied SQUIDs to scanning the human brain, but the system was still rudimentary. Ed Flynn was one of the scientists who turned his hand to coaxing the highly sensitive technology into a shape that was more accurate and efficient. During his tenure at the lab, Ed had built a liquid hydrogen bubble chamber and a wafered solid state neutron detector. He knew how to make things that worked.

Biological sciences at Los Alamos lab had grown up alongside physics. At the time of the Manhattan Project and just after, the life science staff concerned themselves with the health effects of radiation. One of the relatively low-tech experiments in the early days involved exposing physicists to tritium, a radioactive gas, then giving them all the beer they could drink. The predictable and familiar effect, diuresis, washed the tritium out of their bodies, a process the biologists carefully measured.

Current work in biology was a good deal more sophisticated, applying, as Ed did, high-powered instruments of physics to the puzzles inside living cells.

"In nineteen eighty-two, I tried to convince my nuclear physics group leader—a guy I had originally hired, by the way—to go with the SQUID program," Ed said. "He kicked me out of the group, because the work wasn't attracting grant money. My division leader, higher up the ladder, took me in. I worked by myself for a while, then, eventually, a whole new group was created to do biophysics."

Since World War II, the lab administration had maintained a fund to support entrepreneurial staff members who wanted to take off in new directions. In 1983, Ed talked the committee into spending enough money to buy a detector with a single SQUID.

Perhaps no research at Los Alamos had strayed so far from its roots in weapons science than Flynn's brain-mapping project. None had set its sights on a problem so inaccessible, so irreducible as the workings of the human brain.

John Hopkins, the weaponeer I met at the IGCC in San Diego, had said to me that the making of the atomic bomb was the one and only case he knew of where nature made it simple to solve a scientific puzzle. "The Manhattan Project scientists were very clever. They didn't make mistakes. They didn't overlook anything. But fission came easy. It was unique. Take a problem like getting energy

VA HOSPITAL, ALBUQUERQUE

from controlled nuclear fusion—you solve one problem and ten more crop up."

After the success of the Manhattan Project, there arose an expectation in the public mind that if you put enough money into a scientific question, you got an answer out. The brain was certainly not going to yield itself up so easily. It was not even self-evident what questions to ask.

ED FLYNN WAS BALD, SLOW-TALKING, AND RUBBER-faced, with a notable pair of high-flying eyebrows. He led me down a flight of echoing concrete stairs in the lab's Life Science Building to the basement laboratory where a biomagnetometer was set up. This room was Ed's first contribution to the science. It had been sheathed with various metals—aluminum, and a nickel-copper-iron alloy—to eliminate as many outside magnetic signals as possible. "Put a compass inside this room and it wouldn't know what to do," Ed said. Nothing inside the room could be metal. The table where human subjects stretched out for tests was a blocky, unfinished wooden bench piled with fuzzy blankets. The experiments run there inside the cubicle gathered fundamental brain-mapping information: which regions of the brain, for instance, fire up, and for how long, in response to hearing a long, pure tone. Every test also served to size up the equipment and continually refine the technology.

The whole system of acquiring the brain signals and using mathematical algorithms to deduce their sources is called magnetoencephalography, or MEG for short. After designing the shielded room, Ed put together a computer system to process the raw data coming from the sensor.

The earliest instruments were cumbersome. SQUIDs were costly, the necessary sensors custom-made, and the whole package had to operate at very low temperatures, which meant burying everything inside a Dewar flask of liquid helium. With only a single detector, scanning a brain was excruciatingly slow.

Ten years into the project, Ed's group had expanded to include more than a dozen scientists. He had helped to attract enough money from other agencies to build a clinical offshoot at the Veterans Hospital in Albuquerque, where a $2.5 million machine was up and running. And Ed collaborated in inventing a brand-new, patented detector, one that was self-shielding and so saved the cost of building an elaborate and expensive room.

In 1989, the Physics Division brought in a Yale research professor, C. C. (Chris) Wood, to lead the newly formed Biophysics Group. While Ed whittled and wired new devices into being, Chris ran the store, a job that meant juggling the demands of both science and bureaucracy. Besides developing MEG, Chris's group did research in a number of other areas, including x-ray crystallography of proteins.

Chris was built along the lines of a sequoia: sturdy, and taller than everybody else. If a tree could speak, it would be in Chris's baritone. He had dark hair and a thick, untrimmed beard. His keen hazel eyes were startling: when he turned his gaze my way, I automatically ran a check to make sure I wasn't doing anything foolish. Unlike any other lab office I had visited, Chris's took into account niceties beyond pure utility. There were even pictures on the wall.

"Nearly all of what we do is understanding the instrument," Chris said. The instrument fed out data that had to be transformed into a multidimensional model of brain activity. The research team worked on refining the mathematics that estimated just which regions in the brain generated the measured magnetic fields.

The team also devised experiments to reveal what it took to "light up" those brain regions, experiments that flashed images at volunteer subjects, or poked them, or asked them to listen to words that began with the letter "k."

"Mainly, we have used MEG to locate sensorimotor regions in the brain," Chris said, meaning those sites on the cortex where groups of neurons correspond to parts of the body that receive sensations—sound, light, touch—and parts of the body that require muscle control, like elbows, toes, and eyelids.

"It's important to keep these brain-mapping studies in perspective. The scientific and technical problems facing the brain-mapping researcher are daunting enough, but they are only a small part of the larger problem of trying to understand how the brain makes the mind.

"We know the brain makes use of topographic principles, connections based on proximity," Chris went on. "There is an orderly relationship between the position of a stimulus in the visual field and the position of activation on the visual cortex. It is as if spatial relation was projected on to a rubber sheet. The sheet is slightly stretched and distorted, but the arrangement across the surface corresponds to the outside world.

"And we know that in most animals, the amount of brain tissue devoted to a body part often relates to function: relatively large areas of the human brain represent the face and mouth; whereas mice have large areas devoted to whiskers."

But a map should not be mistaken for understanding. "Parts must be studied, but each part must be seen as functioning inside a system to have meaning," Chris said. "Ignore the whole system and you can easily fall for the grasshopper fallacy: A scientist wants to locate a grasshopper's organs of hearing, so she teaches a grasshopper to hop at the sound of a click. Then she snips off the insect's right front leg—the grasshopper still hops at the sound of the click. The scientist snips off the left front leg—same result, until at last the back left leg is cut off and the grasshopper no longer hops when it hears a click. The scientist says, aha, the grasshopper's ears are located in the back left leg!"

Chris emphasized what could and could not be done with brain mapping. "Thought and other cognitive processes are ultimately mediated by the anatomy of the brain, but simply knowing anatomy—a difficult challenge in itself—is not enough. In addition, the cartographers of the mind/brain want to know exactly how the structure and organization of the brain cells and systems of brain cells are capable of mediating all of the complex functions we humans perform—including this conversation."

Chris was often at the lab before dawn and long after dark, but he also knew how to enjoy himself when he was at large. When we got together, he preferred to endure the interview over a beer or with a glass of wine—a carefully chosen glass of wine.

He took great care to impress upon me how modest a beginning the MEG technology was in our efforts to understand what was going on inside our heads. The utter complexity of the problem intrigued Chris, who was aware that progress is not usually made by great leaps but by an accumulation of very small, meticulous steps. "If a scientist's goal was the big breakthough, the ultimate synthesis, then none of us would ever be happy. Science is a mix of hunches, observations, theories, and stories—not fixed bodies of facts. Science is mostly process, and scientists like that process."

VA HOSPITAL, ALBUQUERQUE

•

THE HUMAN BRAIN IS ABOUT THE SIZE OF YOUR TWO FISTS held together at the heel of the hands. Its thickly grooved surface is a spongy layer of gray matter, the cortex. Deeper inside are distinct structures, large and small, such as the cerebellum, a substantial bulge tucked under the back of the brain; two small ovals called the thalamus; the tiny hypothalamus, about the size of a pea; and the medulla, no more than a knob at the top of the spinal cord. These and the brain's other components are not isolated. They connect with one another; they connect to the body and through the body to the outside world. Out of this network arise phenomena as unalike as respiration, balance, vision, grief, golf swings, the smell of an orange, and the tune to "Happy Birthday."

"The mystery of consciousness has been with us from the beginning of recorded history," Chris said. "But the idea that the brain is not just undifferentiated soup took hold in the late eighteenth century." A scientist named Gall dissected the brains of individuals whose characteristics he had catalogued before death. He observed a connection between speech and the brain's frontal lobe. Gall, unfortunately, went off the deep end and assumed that bumps on the skull mirrored enlarged functional areas in the brain. His crackpot theory of phrenology specified an anatomy of hope, destructiveness, and other peculiar categories of thought.

Defining what goes on in the brain more precisely than Gall did is still a leap of faith for neuroscientists. "What, where, when, and how—these are the questions that cover the terrain of modern cognitive science," Chris said. "*What* does vision do? Among other things, it separates objects, it controls the hand's reach to grab the object.

To design an experiment, you take your best guess in defining 'what.'

"Where an event happens in the brain—which brain structures, which regions of the brain are involved in a particular sensory, motor, or cognitive process—is another question, but it does not necessarily fit neatly to the first, in a one-to-one relationship." A given visual perception, a seemingly single *what*, can activate a score of sites across the visual cortex.

"Then you can ask, 'How do various physical entities in the brain coordinate to accomplish the *what*, the visual task of picking out the figure of a cow from the background of a field of grass, for instance?' The *how* includes the character and behavior of individual neurons and the way each fits into a larger pattern of synaptic connections."

This is all much harder than it looks. The first step, segregating out a definition of a specific activity in the brain, a *what*, for purposes of examining it through experimentation, defies our intuitive feelings about the mind.

From the inside of consciousness, where we all reside, experience is seamless. We are unaware of any interceding steps between casting our eyes upon a pasture and distinguishing a cow, a tree, a fence, and the color of the grass. We sense none of the swift neurological transmissions between seeing the cow and recalling the word "cow." But there has long been evidence that thought and perception fragment into unexpected subassemblies inside the labyrinth of the brain.

Until a few years ago, a great deal of the science of higher mental functions like learning, memory, speech production, and so on, was drawn from observing people whose brains had been damaged by trauma or by disease. Lesions on the brain can interrupt the system so that, for example, a person could recognize a cow, looking at it from the side view, but would not be able to make sense of it from, say, overhead. Patients with injury to a certain area on the brain's left hemisphere cannot recall proper nouns like "Empire State Building," but easily remember almost all common nouns. Cases like these call into question any simple guesses as to how memory or perception organize themselves and what their parts are.

Blindness denial, or Anton's syndrome, is a consequence of brain damage that demonstrates how extremely odd divisions of labor in the cortex can be. Blindness denial is a circumstance in which a person cannot see at all but is completely unaware of the fact. The person's behavior can be normal in every other respect, yet, if asked what color the doctor's tie is, will answer, "red" despite the fact the doctor wears no tie at all. After stumbling over a chair, the patient might complain that the lighting is too dim. This syndrome and others like it frustrate otherwise plausible theories of what the brain is doing as it goes about its business.

Building a picture of mental functions based on trauma to the brain is not a thoroughly reliable method. A stroke or accidental lesion will damage tissue randomly across disparate areas of the brain. An injured brain is by definition different from one that is working normally.

Looking at the active, living, normal human brain—determining the geography of brain function, the *where*—reinforces and, in some cases, redefines the problem of *what* is going on.

Until recently there were very few ways to inspect the living anatomy of the mind. One way to do it was to study the exposed brain of a patient already undergoing surgery. Brain tissue feels no pain. Researchers can stimulate different points on the cortex and evaluate the effect on the subject who is awake but comfortable because of local anesthetic. Probing in one place or another on the brain could evoke a memory or, in at least one case, produce a realistic but false memory. A bilingual subject stored the Greek names of things in one square centimeter and their English names in another. In fact, areas involved in language were scattered all over the cortex and varied quite a bit from individual to individual.

Invading the living brain with an electrode was, in these cases, beneficial: the surgeon knew how to leave all the Greek nouns intact. The same kind of meddling with a healthy brain was unthinkable. It was going to require extremely sophisticated machinery to chart brains at work without relying on injury or trespass.

A PHYSICIAN CAN PARK A STETHOSCOPE ON A PATIENT'S chest and hear the gushing rhythms of the heart an inch away. The brain, wobbly and vulnerable as a jellyfish, is locked inside a strongbox of bone. You can't see it or hear it or feel it from the outside.

MEG was one of a few technologies developed in the past decade or two that could capture a minutely accurate picture of the brain through the skull's defensive armor.

Several times a week, Chris drove down to Albuquerque, where a MEG facility was set up to run tests on patients at the VA hospital.

In the VA clinic, patients scheduled to undergo brain surgery, for removal of a tumor, for instance, can get the benefit of MEG's careful mapping of sensorimotor areas. MEG was also being applied to patients with a variety of neurologic and psychiatric disorders, including epilepsy, stroke, and schizophrenia.

VA HOSPITAL, ALBUQUERQUE

I decided to join Chris's subjects in Albuquerque and volunteered my own head for a MEG experiment. A grizzled old man in a wheelchair had stationed himself just inside the entrance of the hospital. He was wearing a plaid bathrobe. A blanket barely covered the raw stumps of his amputated legs. Every time the door slid open and shut, it drew in the warm spring air. This was his front porch.

I followed a corridor that shunted me through oblique angles and blind alleys until I lost all sense of direction. Every hospital has these hallways as a lesson that you are no longer in a familiar world.

The New Mexico Institute of Neuroimaging was crisp and impeccable and punctual. A couple of technicians measured my head and inked landmark dots in a few places. They strapped a few bits of plastic paraphernalia on my hand, a device that beat a soft tattoo on the ends of my fingers. The two of them stretched me out on a platform, pointed a giant upside-down Thermos bottle at the side of my skull, and told me to keep still. They left me alone in a room with the biomagnetometer, but kept in touch over a speaker system. "We are going to start poking your thumb now," said a gentle voice. "It would help if you would concentrate on your thumb while we are doing it." We went through the combinations: left thumb, right side of the brain; left thumb, left side; right thumb, both sides; and then on to my index fingers. Thirty-seven SQUIDs and sensors were discovering where my digits registered in my cortex. The pokes were repeated several hundred times so the finger-related neural pops could be distinguished from the general goings-on, like my wondering what I was going to have for lunch. I lay there, totally nonmagnetic: stripped of my glasses, hairpins, snaps and zippers, in a cotton hospital gown, and tried to think about my thumbs.

The next step was to take a portrait of my whole brain using Magnetic Resonance Imaging. The electronic data gathered from the MEG test would be displayed as a few vivid white dots on a ghostly profile of my gray matter.

Weeks had passed between these intercranial snapshots, and now I was considerably less excited the second time around. Sliding inside the MRI machine was like being dropped headfirst into a test tube. It sounded like a thumping, roaring, grating bulldozer gone mad. Between the time of the MEG test and the MRI, I had broken my wrist and had surgery to put it back in order. My arm still hurt and I was not happy once again to be helpless on my back having unfamiliar activities performed on my person.

The MRI produced a beautifully detailed but static picture of my brain. A radiologist looking at the x-ray–like image would not be able to tell if I had been alive or dead during the procedure.

MEG was one of the few new systems that recorded the brain in action. Another, PET—positron emission tomography—created lovely, multicolored pictures showing variations in blood flow to active regions in the brain. MRI could be combined with other technologies to register a dynamic image of the brain. Each saw the brain slightly differently, with different levels of fineness. MEG was unique among brain-imaging technology: besides triangulating in on *where* a cluster of nerve cells was firing, it could trace tiny volleys of electrical motion millisecond by millisecond. Timing was as critical as topography. It takes a person about five hundred milliseconds to respond in a visual recognition test. If you are trying to reconstruct what has happened in the brain, you know, in this case, only a few hundred neuronal events had time to occur.

"We would prefer to look at each neuron individually,

of course. An exhaustive description of all the neurons would be a starting point for theories of mind," Chris said. "Now what we learn about the way nerve cells group themselves is determined by methodology, by the level of resolution. Every gizmo gives a biased, incomplete picture of what you're looking at." Each separate neuron that fires simultaneously within a focus of activity recorded by these instruments might be performing its own distinct task. "We learn the most by concatenating images from all kinds of technology."

What brain imaging had already revealed showed, as Chris said, "Our seat-of-the-pants ideas need lots of amendment."

A region at the back of the brain fired when a subject looked at a written noun. Hearing the same word spoken lit up an entirely different group of neurons on the side of the brain. It came as a surprise to discover that nerve cells that are activated when a word is heard do *not* fire when the subject him- or herself speaks. Apparently we don't listen to ourselves talk.

When a subject was asked to read a noun—"cup," for instance—and then think of and speak an appropriate verb—"drink"—two more distinct brain areas came into play. After a quarter hour of practice in generating these verbs, the subject's brain quieted down and reverted to using only that small part of the brain needed for the original task of reading out loud. When things got easier with practice, there was an analogous change in brain function.

Some of the team's discoveries seemed natural: the same brain area lit up both when a subject thought about a telephone and when a subject actually saw a telephone. Some of their discoveries seemed very odd: using SQUIDs, researchers determined that the brain reacted to loud sounds in one region and quiet sounds in another.

Scientists expect that all the scraps of data gathered from experiments with brain imaging will eventually cohere into a global map of what happens where. But the journey between a brain's geography and the gestalt of consciousness is not on a straight road. "There is a built-in constraint on brain mapping," Chris said. "The more abstract the functions, the less localized."

Brain mapping was not the only field in neuroscience. Researchers were studying the brain on many different scales: the architecture of different kinds of neurons in the brain, their intricate chemical language, and their telecommunications hardware. From teasing out individual molecules to sketching the brain's gross anatomy, all efforts attempt to break the system down to understandable units, to name them and their functions.

The science of the brain is young. One of the first steps in the science of biology was taxonomy, the giving of names. Classifying warthogs and catfish and maidenhair ferns, analyzing what made them separate species, made it possible to draw them all back together again under the umbrella of a single theory. It was the categories of taxonomy that fed Darwin's imagination and allowed him to recognize the process, evolution, that unified the astonishing plurality of life. Taxonomists could not have guessed just how far their work would travel, but you don't begin an act of discovery and expect to stay in the same place.

I MET WALLY BRAUN IN AUGUST OF 1993, THREE WEEKS after a tumor the size of a pecan was removed from his brain. Wally was so skinny the pant legs of his hospital greens drooped and dragged behind his sneakers. He

VA HOSPITAL, ALBUQUERQUE

looked over his shoulder at the extra yardage and asked the nurse, "Will they pay me for mopping the floor?"

Wally was a stringy, weather-beaten fifty-eight-year-old. He was back at the VA, for follow-up MEG tests.

"He was always a guy who laughed a lot," Virginia Dontje told me. Virginia was Wally's girlfriend, a tiny, agile woman with an electrifying white-blond crewcut. "And suddenly he was angry at everything. It got so bad that one day he walked right into traffic shaking his fist. He kept thinking things were where they weren't. He would pull a chair way out and sit down two feet away from the dining table."

Wally got worse. One day he fell down, partially paralyzed on his left side. He had for a while been under treatment for lung cancer. The first doctor he saw thought the new symptoms were from a stroke. Eventually Wally found his way to the VA hospital in Albuquerque, where the tumor was discovered.

Wally was a tough, tireless self-starter. He had built six World War I airplanes and flown them all. He had built a log house in the Black Hills, cutting all the logs by hand. He had been a claims adjuster, and had owned a couple of western stores. Now Wally and Virginia ran a chrome-plating business out of their home. They loved working on old cars. Wally and Virginia used to roar into Las Cruces on their motorcycle on the way to his chemotherapy sessions.

"I don't remember much of what happened for about six weeks—I don't want to remember," Wally said. "I know that while I was in the hospital before the operation, I was convinced that I was sleeping in a tent alongside Interstate 40 just outside of Gallup. I knew how to push the button to call a nurse, but all the time I could hear the cars go by and the wind blowing the tent flap."

Virginia remembered, "He v as partly blind, but he doesn't remember that either. They gave him some steroid treatment and came and asked him how he was doing and he said, 'It's a miracle. I can hear out of my left eye!' We couldn't tell if he was joking."

Virginia saw her chance to exploit Wally's general confusion. "While he was waiting for surgery, we finally got him to stop smoking. We used nicotine patches on him and gave him straws and told him they were cigarettes."

Wally had ended up at a hospital equipped with the kind of advanced technology, in this case MEG, that made impossible surgery possible.

Dr. William Orrison is a neuroradiologist, the director of the hospital's New Mexico Institute of Neuroradiology, and a collaborator in the Los Alamos team's MEG research. "A year ago," he said, "we might not have done surgery on Wally. He came in with one side of his body extremely weak. His CAT scan and MRI indicated that a tumor was the likely cause. Then comes the next question: Could surgery be done without leaving him paralyzed? Because we had MEG, we could localize function all around the tumor and we could see that there was room to avoid those motor areas. We went in knowing what not to touch. The neurosurgeon knew just how much normal tissue he could take to make sure he was getting as much of the tumor as possible.

"We have used MEG to prepare for fifty operations in the last eighteen months. Nobody woke up worse off than they were going in. Usually, after neurosurgery, several percent of the patients emerge in worse shape."

Wally's sessions under the Big Thermos had been very like my own. "We would ask him to move his hand, or we would apply a puff of air to his fingertips."

The neurosurgeon, guided by dots of light depicted on

the image of Wally's brain, could check the accuracy of MEG's placement. "He can touch the brain at this point and the foot flips up. Touch it one centimeter away and nothing happens."

Bill had been practicing medicine and teaching neuroradiology since the mid-seventies. "Fifteen years ago we were stumped every day. A standard neurological exam—looking at reflexes, asking questions—helped us zero in on an area of the brain. A lightninglike onset of symptoms usually meant a stroke, and gradual symptoms were typical of infection or encephalitis or a tumor. Our diagnoses were only right about seventy-five or eighty-five percent of the time. A huge tumor in the frontal lobe can cause zero symptoms. A tiny tumor deep in the brain can be devastating. Without imaging, we were always a little in the dark.

"Using MEG is less subjective than visually reading a CAT scan. Looking at a patient can introduce bias into reading pictures. If the patient has a deficit on the right side, for instance, you look at the brain scan until you find something where you want it to be. And you don't look for or see abnormalities in unexpected areas."

MEG can do more than evaluate a patient before surgery. "With MEG, we should be able to track the course of disease, and track the process of healing," Bill said. "Here is an example—we know that a small amount of amphetamine can help in recovery from head injury. With MEG, we might track function to quantify the effects of the drug and test when to give it, how much to give and for how long for the best results.

"We could evaluate the brain function of a stroke patient before and after therapy to see in black and white if their brain is in fact responding."

Wally's horseshoe scar, behind his right ear, was about

twelve inches long. "While the doctor was sewing him up," Virginia said, "Wally moved his left arm. They told me the doctor was so excited he hollered, 'Great!' "

"My memory came back instantly, right there in Intensive Care," Wally said.

By the time I met Wally, he was in fine fettle. Just once he forgot the name of something he was trying to describe.

I had to ask him to sign a consent form allowing me to write about his medical history. "No problem," Wally said. "I've signed for everything but permission to be violated."

"That's what this is," I said.

"Well, then, can I predate it and move this thing right along?"

When Wally laughed, his cheeks rose like little red apples.

CHAPTER

6

MARS

> Raids are our agriculture.
> —Bedouin proverb

IT WAS THE DAY BEFORE THE DELTA ROCKET WAS SET FOR launch from Cape Canaveral. Drew, the test engineer, was trying to persuade Bill that there was no need for yet another "final" check of ABE. Drew, as usual, wore red-flowered shorts, a threadbare T-shirt, and no shoes at all, but Bill had learned to take him seriously—the boy knew rockets.

"One spark, Bill, one little electric blue discharge and the whole thing blows up." Drew snapped his fingers. They had already run it through its paces three times over the course of plugging the instrument into the launch vehicle. "It's like raising a child," Drew said. "A time comes when you have to back off and say, 'I've done all I can do.' "

That was enough to convince Bill. Drew was no older

than his daughter and, for once, clearly did not know what the hell he was talking about. "Test it again," he said.

Bill Feldman, a physicist at Los Alamos, had spent three years on this satellite. The instrument was called ABE, the Army Background Experiment, and it contained Bill's invention, a boron-loaded plastic scintillator. The scintillator was a sensor, part of a Strategic Defense Initiative program that was going to look for nuclear warheads in space. Bill, a mild, gray-bearded, good-humored man, would not relent and go back to the Holiday Inn until he checked ABE out.

The phone rang in the control room and Drew answered. He listened for a moment before breathing out an amazed, "Goddam." A mechanic tightening the last screws in the nose cone had dropped a wrench. Hand tools were always tied to a worker's wrist, but this had been a hinged device for going around corners: one part popped free, fell, bounced off a solar cell and crash-landed right on top of ABE. There was no more argument—they were going to have to run a test just to see if the instrument was still alive and well.

By evening, they knew ABE had survived the accident. It was a lush, tropical night and the lead man on the engineering team invited Bill to come with him to see the world from the tip of the Delta rocket.

From a distance, the gantry looked like a delicate web, but it was solid as a skyscraper. The men rode up most of the seventeen stories in an elevator, then climbed a final few flights of stairs up through the steel grid that encased the Delta.

The rocket's nose cone, the part that carried the payload, was inside a brilliantly lit, clean room that enclosed the top of the gantry. ABE was a boxy metal butterfly, mounted on wings folded to fit into this ten-foot cocoon. Bill and Drew donned surgical scrubs and caps before entering the room: ABE's solar panels, the source of energy after deployment, had to be kept free of dust and hair.

Bill looked ABE over for the last time.

Back at Los Alamos, months later, Bill recalled the launch. "You start to feel affection for the machine. It's unique, it's handmade, it's part of your life, and it's about to be brutalized. You set off what amounts to a bomb underneath it and push it up through the air where it rattles around like you bounced it down the side of a mountain. I felt like I was sending my toddler into traffic."

ABE survived. The rocket lifted into space, and, inside the nose cone, a tiny explosive charge detonated to open the fairing like the halves of a steamed clam. ABE popped out on a spring—a million-dollar jack-in-the-box—spread its wings, and flew.

After seven months, ABE was sending home data that agreed with all previous calculations. The box was working.

"You never know if a sensor will degrade in orbit," Bill said. "Out in space, there are solar flares, galactic cosmic rays, all sorts of radiation. Lots of substances will change color or fog up. Have you ever seen a really old bottle turned blue or green? That's from natural radiation from cosmic rays over a very long period of time. Some substances get brittle or crack from radiation and from the temperature changes that occur when a satellite goes in and out of a planet's optical shadow. We test for these things before we send equipment up, but you can't know for sure. Once you're out there, solar protons can give false instructions to your electronics—they're like gremlins. So much can go wrong."

Interplanetary space is a hostile place. On the one hand, it is nearly empty and, at 270° Centigrade below freezing, profoundly frigid. On the other hand, it seethes with wild cosmic winds: brutally vigorous particles and fields of force that bend and snap in endless flux, forces as tangible as a wave cracking down a bullwhip.

Bill's scintillator passed its survival test. Its next mission was a trip to Mars to see what it could find there.

THE RUSSIANS LAUNCHED SPUTNIK INTO ORBIT IN 1957, throwing a scare into everybody. Sputnik achieved, psychologically if not practically, the strategic high ground. The day after the satellite went up, Eisenhower doubled a budget request for ICBMs from forty to eighty. Money from federal coffers soon flowed freely into university research programs, an investment in technological catch-up. A new agency, the National Aeronautics and Space Administration set out to stake a claim in the cosmos. The Arms Race was from then on accompanied by its shadow, the Space Race. But there would be more than muscle-flexing spy satellites in space; there would be mystery, heroism, and frontier adventure: a man's footprints on the moon, and robots on Mars.

The laboratory at Los Alamos quite naturally joined the Space Race. In the mid-1960s, they built satellites in a program to detect nuclear testing above the atmosphere in case the Russians were devious enough to try it. When, thirty years later, Ronald Reagan set Star Wars in motion, one of the lab's assignments was figuring out how to detect an *unexploded* nuclear weapon in space. Bill's ABE was an offshoot of Star Wars, the Strategic Defense Initiative. It occurred to Feldman that one way to determine whether an object shooting through space is a loaded weapon or a decoy was to look for neutrons. "Where there is fissile material there are free neutrons flying around," Bill said.

ABE was a first step, a device that could take a census of the natural neutron population, the neutrons that were there before a weapon came on the scene. Eventually, the experiment would have a military application, but for the moment it was just basic science, surveying space to see what was out there.

"The bean pushers would prefer to just get the job done and forget about scientific context, but they are amenable to reason," Bill said. "At the policy level, the military is like all institutions, there are some smart people, some dumb ones, even some altruists. There are military people smart enough to realize science is a very inefficient process moved along by curiosity and stamina and that scientists need a kind of nurturing." So Bill's neutron detector snuck into the Star Wars budget, but he quickly saw other suitable employment for ABE: searching for water on Mars.

Bill was a slight man. He had a handsome head—high brow, blade nose, curly gray hair—but narrow shoulders. He was thin as a seedling, except for his thick, blunt, ditchdigger hands. His smile was a picture of dimpled geniality, and his voice cracked like an adolescent's. "I went into science because I couldn't fight and I was no good at sports," he said. "Bronx Science was a better choice than the blackboard jungle school that was the alternative."

Bill grew up on Manhattan's West Side, the son of a physician. He studied at MIT, then moved to Stanford for his doctorate in nuclear structure. He met his wife, Margrethe, a visitor from Denmark, while he was in California. Their courtship began with a motorcycle tour of the coast. Bill joined the Army after getting his Ph.D. "I wanted to repay my dad for fighting World War II."

His tour of duty took him to Ames Research Center, a NASA laboratory, where he worked as a member of the Space Plasma Group. He signed on at the national lab in 1971 in part because Los Alamos was a quiet mountain town. For the first ten years, he worked on other people's projects. In the early eighties, Bill struck out on his own to create this box that could measure neutrons.

"Looking at a wall, you look at the uppermost angstroms," Bill explained. "What you can see is controlled by the wavelength of photons. That is the skin depth as far as the eye is concerned. But you can interrogate material to different depths."

Neutrons pass easily through something like a rock. They will carom around inside, slowing a little bit, and taking on the "temperature" of the material.

"Neutrons end up acting like a thermometer," Bill said. "On Mars, you can determine the properties of the upper meter of the surface—a skin depth of around three feet—to find out what's there. The neutrons go in, acquire the spectrum characteristic of that material, whether it's iron or oxygen or water, and then they come out again."

As the Cold War petered out, Star Wars lost its funding, but NASA found room for Bill's detector on Mars Observer, a spacecraft scheduled for launch in 1992. It would be the first U.S. mission to Mars in seventeen years.

IN JUNE OF 1991, FIFTEEN MONTHS BEFORE LAUNCH, Bill's scintillator was in the final stages of assembly at Martin Marietta in Denver. To save room on the spacecraft, the instrument was piggybacked with another, a gamma ray spectrometer. The engineers were building the whole package on a frenetic, nightmare timetable, racing toward the launch date.

Bill was going to test the scintillator using an artificial source of neutrons. The Source: this is a polite way of labeling a highly radioactive lump of americium boron. Americium is a metal manufactured by bombarding plutonium with fast neutrons. The whole package was hotter than a pistol and Bill had filled out a few pounds of government forms to arrange its passage across state lines. He drove it up from Los Alamos in the back of a Chevy van. He would use the Source to flood his detector with a known quantity of neutrons to see how well it counted.

At Martin Marietta, a number of interconnected laboratories had been given over to the fabrication of the gamma-ray spectrometer/boron scintillator instrument. The instrument itself was out of sight, inside a vacuum tank cooled by liquid nitrogen. Inches of soft ice caked the metal barrel, which exhaled slow clouds of condensation. Dozens of odd-sized cables snaked out of tube ports in the tank, hooking the instrument to electronic monitors stacked two deep on shelves and table tops all over the room.

In the lab next door, the Source sat, hermetically sealed in stainless steel vials, stored in a metal box filled with wax. Malleable sheets of lead and cadmium soft enough to bend in your fingers had been loosely shaped around the box. The whole arrangement rested on a wheeled table in an alcove taped off at waist height by yellow scene-of-the-crime tape.

"Packaged the way it is," Bill said, "it's safe enough to carry in your arms."

Before he could move the Source, Bill had to notify an occupational safety officer and the Martin Marietta lab director for authorization. When he got the go-ahead, it was just a matter of trundling the table next door into position next to Bill's scintillator. I offered to help steer—if Bill said there were no death rays, I was going to take him at his word.

We removed the yellow tape and walked up to the box. Bill held out his arm and said, "Don't touch the lead shield if you can help it. Those heavy metals can kill you."

Bill took the pellet of americium boron out of its steel jacket and inserted it into a solid block of graphite coated with an inch of polyethylene. Neutrons, imperceptible to us, streamed from the Source and into the scintillator. Across the room, far enough away to be safe from the radiation, Bill bent over a viridescent screen that would show the raw numbers translated into a simple line graph. Bill knew the shape he wanted to see, a portrait of the neutron flux whose accuracy meant the detector was in perfect working order.

"That's it!" He leaped up and started to pace back and forth in front of the monitor, ebullient. "This is beautiful. Thank heaven."

Inside the frozen tank, Bill's contraption was seeing neutrons and telling the story. The test's success was a boarding pass to Mars.

I couldn't travel to Colorado in the van with the Source—it was forbidden—so I had driven up alone on the interstate out of Santa Fe to watch Bill at work. At the start, I-25 dipped south through sharp redstone mesas, foothills of the Rocky Mountains. Right before Las Vegas, New Mexico, the road swung wide to head north, skirting the edge of the Great Plains: the horizon went out what seemed a thousand miles flat. In June, spring grass stretched to the end of the earth; thunderheads slid along the top of the sky. An hour into the trip, I could see worn-down cones of old volcanoes like a line of tents crossing the prairie from Raton to the Oklahoma border. The local radio station was broadcasting news of lost dogs: a yellow Lab was missing, and a female Samoyed, black and white. At that moment, it occurred to me to wonder what strange appetite could draw Bill away from this world toward the lifeless sands of Mars.

In a Denver steak house the evening of the scintillator's successful dry run, Bill tried to give me an answer. "You look at a Georgia O'Keeffe painting, a flower, and you feel it is beautiful. Only then do you begin to ask why it is so. We all observe nature all the time. Sometimes you see something that completely captures your attention. When you ask yourself what is going on, you go deeper and deeper trying to find out."

That was part of the Space Race, too.

AS IT TURNED OUT, MY VIEW OF MARS WAS EXACTLY backward. The point is not that it is so alien a place. Mars is a compelling destination because it is so like earth. The two planets are twins separated at birth.

Now Mars is frigid, dust-dry, antiseptic. Earth is hot, with a pulsing heart of molten rock, covered with a tumult of biotic exhalations—all that will smooth the sharp edges of geological upheaval. On Mars, monstrous stone features—Olympus Mons, a volcano big enough to blanket New York State and spill over onto Pennsylvania; the Valles Marinaris, a 3,000-mile canyon that slashes across the belly of Mars like a mortal wound—have stood nearly unchanged, dead for eons. But it wasn't always so.

When the United States space probe Mariner 4 turned its camera on Mars in the summer of 1965, it saw a bleak moonscape. The intriguing patterns of light and dark seen through earthbound telescopes did not correspond to any details in the dull, cratered blankness of the Mariner's pictures. Years of eager speculation about abundant life on Mars were doused in an instant: the patches weren't canals; they weren't vegetation; they weren't anything. A few years later, Mariners 6 and 7 flew by and caught glimpses of slightly more interesting terrain—vast, smooth plains and swaths of jumbled rock. Instruments on the spacecraft measured extremely cold temperatures on the surface and only a faint halo of atmosphere, mostly carbon dioxide.

In 1972 Mariner 9 visited Mars and stayed awhile. By sheer chance, the earlier explorers had seen only the planet's moribund southern hemisphere. Number 9 revealed the northern regions of Mars, a place of epic volcanoes and canyonlands, signs something more dynamic than dead rock was to be found. Most important were the channels carved so much like earth's own streambeds and flood plains. These dry tracks of bygone rivers forced the conclusion that Mars had once been warm and wet, with water flowing freely across its face.

Physics is the same everywhere in the universe—that is axiomatic. Nevertheless, in the billions of years since the Big Bang, the clean, simple laws of physics have fractured into localized complications of prodigious variety. A glance around the cosmos tells you that atoms have put themselves together in an astounding number of ways. There are black holes, exploding supernovae, brown dwarfs, quasars, invisible dark matter—it's a madhouse. From the vantage point of the smooth, symmetrical bubble of raging-hot quarks that marked the beginning of time, it seems absurd that carbon and hydrogen and oxygen would get up on their hind legs and start looking for answers. Yet we know for sure that it happened right here. The big question is, did it happen anywhere else? Are we marooned inside these hundred billion galaxies? Was this whole extravaganza just for our benefit?

On the theory that you do what you can with what you've got, planetary research took the shortest route to answering the big question: look for life next door to where you know it already exists. News that Mars, in its youth, was warm and wet excited great hope in the 1970s. The United States sent Viking I and Viking II to drop robots right onto the surface of the planet to find living organisms.

Warm and wet matters because life cannot spring naked into the universe. It needs a womb of water and atmospheric gases to grow in. Almost everybody agrees that water is absolutely essential to life. "Organic compounds need a solvent to react to one another and to inorganic compounds," Bill said. "Water is a cosmically abundant and efficient solvent. Salts dissolve in water—sodium chloride and hydrogen sulfide—to form a medium in which chemistry can proceed rapidly—you get ions searching around for attachments. Water is also a temperature buffer. Too much heat can break up chemical bonds; too little and nothing happens. There have to be a lot of chemical encounters to make the correct combination for life to happen."

Water, in the form of a liquid brine, busy with mating molecules, can only occur in that very narrow slot around room temperature. If it was indeed flowing water and not liquid sulfur or some other oddball agent that carved out the dry riverbeds on Mars, then that is a clue to temperatures in the past. Without the weight of a reasonable atmosphere, water instantly evaporates into a gas. So, when you assume liquid water on Mars, you must also assume a denser atmosphere. Mariner 9 sent back portraits of gargantuan volcanoes, another piece in the puzzle. Volcanoes act like planetary bellows, gushing gases from the interior's superheated rock out to the surface to build an atmosphere of carbon dioxide. A carbon dioxide atmosphere performs all kinds of services necessary to nascent life: it holds heat from the sun, keeps water in a liquid state, and also shields burgeoning molecules against ultraviolet light. Ultraviolet radiation is so energetic that it easily tears apart the large, complex, loosely bound molecules that are the basis of life.

The ancient tracks of moving water on Mars can be approximately dated with a painstaking count of the impact craters that overlay the channels: How long have they been exposed to the somewhat predictable barrage of meteorites? It appears that water ran free on Mars around three and a half or four billion years ago.

Pull all the evidence from Mariner 9 together and it suggests that once upon a time, Mars had been moist and temperate, strikingly similar to earth three and a half billion years ago at the very moment life emerged here. The best guess is that our own primal ancestors mastered the trick of locking themselves together in long, self-replicating chains of chemicals sometime in the passage of 300 million years. That is a short spell as planetary time goes, and probably no less than the balmy years on Mars. Things were looking good for extraterrestrial life.

The Viking missions to Mars were fitted out with instruments for detecting microbes that might inhabit the desert dust and for tasting the elements in the salmon pink sky. The lander from Viking I scooped up handfuls of the rock-strewn soil and found it to be mostly silicon and iron oxide: rust that tinted the face of the planet deep red. Sensors discovered traces of nitrogen, an essential component for life, in the atmosphere. But the biological laboratory on board ran test after test without discovering a living thing. The lander from Viking 2, 4,500 miles away, confirmed the findings. The sands of Mars are actually a killing ground. The sun's ultraviolet rays react with the soil, producing caustic chemicals that can burn away any organic compounds that might turn up there. The human race was going to stay lonely for a while longer.

The planet, nonetheless, had a story with a moral to tell. Mars's kinship to earth generated certain questions whose answers redound upon human concerns. Since the two worlds were so alike in the past, understanding the forces that turned Mars cold and dry and sterile might inform us of catastrophes impending in our own future. "How did the atmosphere get lost?" Bill asked. "Where did the water go? Like doctors who go into certain specialties because of their personal fears, scientists perhaps shape these questions out of what scares them."

Mars was a museum that depicted the death of an earthlike planet. Its history, pieced together from all that has been learned read like this: Smaller than earth and farther from the sun, Mars more quickly lost the heat of its internal furnace of molten metals. As the volcanoes quieted, carbon, trapped in stone, no longer outgassed to feed the atmosphere. Mars was only about half of earth's circumference, and about one-tenth of earth's mass. Its weaker gravity may have allowed what atmosphere remained to seep away into interplanetary oblivion. When the planet cooled, water froze. A little bit of it remained in surface ice. Although the polar ice caps on Mars were mainly carbon dioxide, "dry ice," a certain amount of water has been detected there. But there was probably enough water on Mars at one time to cover the whole planet to a depth of thirty to a hundred meters. There is evidence that at least some of it remains, sunk deep in the ground, an inner shell of permafrost beneath the visible layer of Martian soil. Underground ice creeps and flows and softens the terrain above, and a number of features on Mars appear squishy. There are cliff faces with mudslides and meteorite craters, called splosh craters, that look just like cow flop.

Nothing in this chronicle has discouraged the life-seekers. On the contrary, now they know more exactly what to look for. If they discover water under the ground, they will search for the kind of rugged organisms that survive in the lakes of dry valleys of Antarctica. There, lakes exist under ten to twenty meters of ice, staying liquid even at surface temperatures far below freezing; and at the bottom of the lakes, algae grows.

If they locate sediments where water stood billions of years ago, they will hunt there for the fossils of living things that arose in the warm, wet past and left behind an imprint for the future to find.

Mars Observer was going to circle low over Mars, pole to pole, taking photographs and measuring temperature, magnetism, and gravitational fields. The gamma-ray spectrometer could analyze the chemistry of the soil.

Bill's box had a special mission. The neutron detector's special talent was for seeing hydrogen. Neutrons that interact with hydrogen take on a unique temperature. If the scintillator recorded this kind of neutron bouncing up from Mars, it would flag the hard-sought water frozen underground.

The scintillator was also sensitive to carbon. "If there was standing water," Bill explains, "it would have leached carbon dioxide out of the atmosphere. Carbon precipitates out as rocks—like limestone." Beds of carbonate rocks on

Mars would be the fossils of former seas, and perhaps the graveyards of extinct species of bacteria or algae.

The neutron detector can draw a treasure map for future missions: this is where to dig for signs of life.

THERE WERE UPWARD OF TWENTY SCIENTISTS, ENGIneers, and technicians on the team running the combination gamma-ray spectrometer/boron scintillator. The first serious sketches, the first useful equations, went down on paper in 1982. Congress approved funding for the Observer in 1985. After the Challenger explosion, NASA waffled over the plans, then snatched fifty million dollars from the Mars budget to pay for shuttle recovery. Two years and a couple of sophisticated instruments were lost in the bureaucratic swamp.

The launch date was ten years into the project. The trip to Mars would take a year. The Mars Observer would start to send numbers back to earth in fall of 1993. Another two years would have to pass before solid facts could be fashioned out of the raw material of the numbers. Bill was going be fifty-five years old by then. Humility is useful to a planetary physicist; patience is essential.

There were also a few short-term rewards.

To calibrate the instrument, the team needed to duplicate conditions on Mars as nearly as possible. Antarctica is the closest match. In 1990, one of Bill's teammates, Steve Squyres from Cornell University, went to Antarctica to dig up soil samples that were likely to be similar in composition to those on Mars. The soil he wanted was at the bottom of a frozen lake. Steve, a skinny lad, is a professor of astronomy and physics. At a cocktail party for team members, he sipped beer from a plastic cup and, with a nostalgic smile, described scuba diving in Antarctica.

"I had to drop down a shaft through seven feet of ice before I hit water. There is a continuous carpet of blue-green algae growing on the floor of the lake. It produces oxygen that is captured in the ice as columns of bubbles that look like architecture, like the interior of cathedrals. The algae itself gets lifted up from the bottom by the oxygen. It forms giant peaks like stalagmites before it breaks up."

I ask him if he was afraid of the cold and the dark.

"I was only afraid of screwing up the experiment. I had so little time to get the samples—you couldn't stay down long—that I was nervous about making mistakes. I was wishing I could just stay there and look around."

In the Antarctic summer of 1992, it was Bill's chance to go. He had the time of his life.

"They issue you forty-five pounds of clothing and gear. The survival course teaches you how to deal with hypothermia, snow blindness, how to get out of a crevasse. Two of us sawed out blocks of ice and built a snow house—it's like an A-frame over a trench—and spent the night."

Bill loved it. "You can see two twelve-thousand-foot volcanoes from the landing strip, Mount Erebus and Mount Terror. From sixty miles away, you think you can reach out and touch them. What look like chicken scratchings on the slopes are giant cracks.

"The American town, McMurdo, looks like Dogpatch. The average age of the residents is around the mid-twenties. They call scientists beakers. Until you've been there a year, you're a fungie, a fucking new guy.

"Nobody says, 'It can't be done.' We needed fifteen hundred pounds of special dirt, weathered basalt, high in iron. It took a helicopter to get the stuff back to the balloon gondola. Nobody ever asked why we didn't just dig closer. They got the job done."

The gist of the experiment was to hang a ton of Martian-like soil from a helium balloon to which various scientific instruments were affixed. The balloon went above one hundred thousand feet to a layer of atmosphere similar in density to the Martian atmosphere and therefore as naked to cosmic rays. The detectors were supposed to pick up gamma rays and neutrons coming off the soil in a miniature re-creation of what would happen on Mars itself.

Besides Bill, the crew of scientists manning the experiment included Steve Squyres, back for a return trip. The work required combination computer hacker/stevedore skills. The physicists themselves shoveled, sieved, and bagged the fifteen hundred pounds of basalt, but that was not the end of it. The balloon's gondola was inside a structure with lots of computer gear that does not like dust. "So we spent a couple of hours draping the gondola in plastic," Steve reports, "and built what amounted to a giant oxygen tent." They then bucket-brigaded the dirt into the hopper.

After some frantic last-minute jury-rigging of batteries and antennas, the weather turned evil. "The wind was howling, blowing snow along at ankle level like little ghosts," says Steve. "If we had so much as taken the gondola outside, the gusts would have ripped the solar panels off."

When the weather calmed and launch day finally came, all 900 feet of balloon was laid out along the snow, fitted with inflation hoses. Once the helium started pouring in, it took an hour and a half for full inflation. It was pinned down for the process, then finally let loose to rise straight up. According to Steve, "The sound of the balloon rushing through the air is unbelievable—like God sitting down to dinner and ripping his pants."

After a perfect launch, almost everything else went

wrong: computers crashed, the balloon oscillated for reasons unknown, telemetry from the primary experiment fed out data streams of nothing but zeroes. Following a day on which the scientists beamed up desperate messages, every detector except the gamma-ray spectrometer, a duplicate of the Mars Observer instrument, answered. The following days were a matter of waiting for the balloon to drift along an erratic circle around the continent until it could be landed and the equipment recovered in a methodically mad dash on the ice.

Bill was lucky. His scintillator operated exactly as planned and gave him everything he needed.

"Now I can see how the neutron detector responds to known quantities. I have a proxy image of what the background on Mars will look like, so we can now reduce the error in background-to-signal calculations."

Steve's gamma-ray spectrometer had been blind and deaf from day one of the experiment. Steve flew to the balloon's remote landing site to unbolt all the instruments from the gondola and salvage whatever was intact. He knew he was not the only explorer Antarctica had defeated. "One thing I couldn't help thinking about while we were on the ground," he wrote later, "was Scott and his men. Even stumbling around the gondola was tough, because you'd break through the snow to mid-calf about every third or fourth step. What a place to die. There was nothing on the horizon in any direction—just flat and white as far as the eye could see."

LEE GOLDBERG SWEPT HIS ARM OUT TOWARD A QUARTER mile of linoleum-tiled, fluorescent-lit corridor and announced, "It's a dream factory."

Lee was showing me around GE Astro Space, a plant in

Princeton, New Jersey, where all the parts of Mars Observer were being mated into a single flightworthy spacecraft. Lee was a no-frills kind of guy. He was bearded, sandaled, baggy-pantsed, and a little heavier than fashion would allow. His formal job designation was Science Accommodations, Mars Observer Program. That translated to fixer, mediator, general dogsbody—Lee brokered agreements between the demands of the seven scientific teams and the limitations of the "bus" that would ferry them to Mars. The craft could carry only so much weight; it could deliver only so much power; it could perform only a couple of tricks. The scientists always wanted more, and Lee had to negotiate the compromises.

"The craft can be hazardous to itself. I did what I could to get the gamma-ray spectrometer out on a boom, as far away from everything else as possible to minimize the amount of rays coming off the satellite and interfering with the detector," Lee said. "We engineered an extension arm at maximum stress tolerance." But the arm is not long enough to completely satisfy the scientists.

We stood at an interior window just outside the High Bay, a three-story lab where the Observer's scientific instruments were put through their paces. The room was so white it looked refrigerated. The technicians inside were dressed like anesthesiologists. Lee and I squeezed in between six-foot walls of electronics.

"The spacecraft talks to us with microwaves," he said. "We demodulate the beacon to get a series of ones and zeros—computer code. The spacecraft sings a song, sixty-four-thousand bits that repeat like a round, 'I am the Mars Observer Spacecraft. This is where I am, this is what I know about the science on board.'"

He pointed to a component where a blue line jigged

across a small black monitor. "And here is a picture of the song—'This is how I am doing.' A forty-watt signal calling us all the way from Mars."

Walking and talking in double time, Lee led me through a winding series of hallways and lofty laboratories, pointing out various species of gleaming boxes along the way.

"GE mostly puts together weather and communications satellites here."

That about exhausted his interest in anything but the Observer. "A geosynchronous orbiter is like a cockroach—tough, stupid, and if it flips over, it's lost."

The beauty of Lee's spacecraft was worth stopping mid-course to explain. "The Mars Observer will have a horizon sensor and a celestial sensor to orient itself in orbit," he said. "It's fairly smart and autonomous. When the sun is between us, we can't direct it, but it can take care of itself. When it senses something wrong, it 'safes' itself—duck-and-cover mode. It shuts almost everything off and signals, 'Something's wrong; I woke up and didn't know where I was.'"

Our destination was another bay where the body of the spacecraft sat in wait for its scientific instruments. We pulled electrostatically neutral polyester coveralls over our clothes and elasticized caps over our hair. We changed into soft canvas shoes and entered a double-doored anteroom. The floor of the anteroom was lined with sticky tape to strip dirt off our soles. The doors locked for a half minute while fans blew dust and hair off our outer garments.

The Mars Observer perched on a pedestal to the side of a raised platform that filled half the laboratory. A communications satellite rested on the airy blue cushion that carpeted the platform. Shoeless engineers tiptoed around on the cushion and watched their baby unfurl its solar panels, rehearsing for space.

We walked up to the Mars craft where it sat waiting its turn to be put through the paces.

"Well, there is no elegance to the design," Lee pointed out.

Nothing could appear less likely to go airborne: the body of the craft looked like an oversized blender with its casing removed. Take it all in and you felt like you were greeting an extraterrestrial insect, with a fat belly and skinny, elongated arms. When it got to Mars and extended its billboard-sized solar panel, the Mars Observer was going to look as overbalanced as an ant hauling a leaf twice its own size.

"Flight hardware is unmercifully driven toward weight and durability parameters. It's brute fixturing—it will be used only once and there is no time to refine the shapes."

Nevertheless, this device in front of me was on its way to *Mars*, as surely as my next stop was Philadelphia. The slightest touch would contaminate the surface, so I raised my hand and ran it through the air a half inch away from the brassy gold-green thermal wrap that shielded a bundle of cables.

This was as near as I would ever get to space travel. This was as near as any of the scientists who had shaped the box would get. Many of them would make the trip to GE—from Los Alamos, and Tucson, and Ithaca—with little reason except to be in the same room with their baby.

You could pick up a pencil to prod an object, and just as if you had touched it with your fingertip, you would know its shape, texture, density, and weight. When the spacecraft beamed coded messages to its keepers on earth, they would experience Mars in a way that is only a step or two more remote than poking it with a stick—as close as that.

I would meet up again with the Mars Observer a few months later in Cape Canaveral to watch it take off for Mars.

That day, Lee and I stripped off our gear and walked back to his office. The very walls of GE/Astro Space were a monument to the fact that no matter how many brains you bring to bear on a project, you cannot think of every contingency: a path of wide horizontal grooves are cut out of the pilasters that projected a couple of inches into one of the buildings corridors. When the spacecraft was first trundled down the hallway, it was too wide to fit. "We call it," said Lee, pointing out the shoulder-high furrow, "the Mars Observer Memorial Notch."

EVERYTHING ALONG THE MAIN DRAG IN COCOA BEACH looked as if it had taken a beating. Grubby pastel apartment complexes and blowsy motels sat frying in the middle of sun-blasted asphalt. Exhausted strip malls wore For Rent signs around their necks.

Only yards away from the banged-up boulevard, the Atlantic hunched against the Florida coast, a huge, gray presence, quieter than the passing traffic.

My hotel room was right on the beach. Through banks of windows on two sides, I could see the shoreline curve out to a point of land, Cape Canaveral. Even from ten miles away, the silhouette of a launch tower was easy to make out. A few miles north of that, just over the horizon, Mars Observer sat in the nose cone of a Titan 3 rocket.

It was the end of September but the sun was ferocious and the air at midday was swollen and heavy: wet wool. Several hundred people were in town to watch the spacecraft take off for Mars, people from NASA, from the Jet Propulsion Laboratory, from General Electric, from Bechtel National, Inc., from a dozen universities.

The festivities were a week late, delayed on account of rain, a lot of rain—Hurricane Andrew, in fact. Mars Observer had been out on the launch pad when the weather started coming its way. Technicians pumped nitrogen gas into the housing around the satellite to raise the air pressure so dust and rain could not seep in. After Andrew passed, the spacecraft looked as though it had spent the night in an alley. It was covered with paint chips, rust, lint, and paper shreds. The hurricane was blameless; it was dirt in the gas lines that contaminated the craft.

Mars Observer was cleaned up and ready to go again on September 25.

The science teams, along with their spouses and kids and friends, climbed on buses at the Kennedy Space Center headed for the banks of the Banana River, an inland waterway on whose farther shore perched the Titan, a few miles away but plain to see. The bus driver intoned some warnings: in the event of toxic gas, alligators in the water, snakes in the sedge.

The crowd spilled out onto a narrow strip of spongy Bermuda grass between the road and the waterline. A hot-dog stand and portable toilets were already set up, in business for the day. A public address system broadcast the play-by-play from command central. There were one or two hours to kill and the wait developed into a long, skinny picnic with touch football, Frisbees, and napping babies. A couple of dolphins scalloped through the water near shore.

The final countdown was like the ringing of a gong, rhythmic and portentous, and then the Titan's engines thundered. The rocket eased upward on a column of fire, flames with the biting brightness of a welder's torch. The rocket's arc was slow and full of grace. Its great ripping, rumbling noise was as wide and deep as the whole sky. The flight was gorgeous, cutting a passageway into space, but stark, too, like a scalpel slicing flesh.

Everyone got back on the bus and returned to the Center to wait for the first message.

The scientists and engineers gathered in an auditorium, expecting to hear electronic chatter from Mars Observer, gliding now over Africa, now India. An hour passed: nothing. It was not bad news exactly, just no news. The men and women were rigid, restless. They had labored for a decade and expected to spend years more connected to the Observer, probing through the information it would send back to them. A whole life's work lay in the kinship. Watching it glide away, out of sight, they were stunned by the thought that it was gone for good. With one tick of the clock there was silence, with the next, the spacecraft spoke, and a hundred people leaped up and cheered.

In August of the following year, as the spacecraft closed in on Mars, Bill Feldman was at his monitor back at Los Alamos. The first data from satellite was expected to come in about one o'clock in the afternoon. All over the country scientists stood by like a gang of kids clutching their ticket stubs, waiting for the gates of Disneyland to open.

Bill had been talking to his neutron detector for months as it sailed toward Mars. The instrument worked perfectly. Zero hour came and went and there was no signal. "It's foreboding," Bill said.

In the days that followed, Mission Control at the Jet Propulsion Laboratory in Pasadena cajoled the on-board computer with one instruction after another but there was no answer. Somewhere in the twists and turns meant to insert the spacecraft into orbit around Mars, the Observer had slipped away and was gone.

It was a strange sorrow, a distant death. Mars was like a lost sister—if we could only see her face, we would know so much more about our own.

"Standing in the northern hemisphere of Mars," Bill said, "you would recognize the night sky—the Big Dipper, Orion—it would look the same." Millions of miles away, but that close.

With twelve years of his work crippled and marooned, aimless in space, Bill sat in his office at Los Alamos talking about what went wrong. On top of his desk was a milk carton printed with a picture of Mars Observer and the question, "Have You Seen Me?" Bill's office was long and narrow. Braced by two computer terminals, hemmed in by a bank of file cabinets, we held a wake for the missing satellite.

"A long time ago, NASA set themselves on a course they thought would guarantee success. If a technology worked well once, they would stick with it instead of using something untried. They ended up with proven, but klutzy, outdated technology, using add-ons and adaptations when what they could have done was start from scratch and make something simple and new.

"Another problem was that a device as large and amazingly complex as Mars Observer requires extreme diligence to build. You have to test and test and test and probe to make a box like that work. General Electric, operating with a fixed-price contract, had to make a profit and that much vigilance is expensive."

Bill offered me a cup of coffee from the kitchenette in the corridor. Like every piece of equipment on lab property, the refrigerator carried instructions: "For Food Only." There must have been another kitchen for toxic substances.

Bill sidetracked to a new topic, a little antidote for loss. He had just signed on to a new project, a satellite to investigate lightning. "Cosmic rays spark an explosive avalanche of electrons—we know almost nothing about how it happens." The experiment would be part of the lab's nonproliferation program, figuring out how to distinguish a nuclear test's electromagnetic pulse from one generated by a flash of lightning.

In a final reflection on NASA, Bill said, "NASA's money came from politicans who had one agenda—Cold War competition—while the scientists had another. And NASA got big enough so that just perpetuating the institution became important—saving jobs, saving the towns that depended on those jobs.

"Now it's the end of the Cold War and the beginning of an era of budget restraint. We have to go back to the beginning and decide what the reasons are for doing things."

Bill's advice was meant for NASA, but applied equally well to Los Alamos National Laboratory.

CHAPTER

7

OPERATION MORNING LIGHT

> Ethics is allocating resources.
> —David Torney, Theoretical Division,
> Los Alamos National Laboratory

IN THE WINTER DARKNESS OF JANUARY 1978, A SOVIET SPY satellite wobbled out of orbit and fell to earth. The satellite, Cosmos 954, shattered as it hit the atmosphere over Canada's Northwest Territories. It strafed the tundra with debris, laying down a 600-mile-long band of radioactive birdshot from the eastern edge of Great Slave Lake to Baker Lake near Hudson Bay. Cosmos 954 had housed a miniature nuclear reactor fueled with 100 pounds of enriched uranium.

The few residents of Yellowknife who were awake that early saw a score of brilliant lights, white, mingled with oranges and blues: what they thought was a jet on fire. They heard a deep, elongated *whoosh* as the flames streaked down through the black sky.

Eight hours later and thousands of miles to the south, Dr. Carl Henry boarded a C-141 transport plane. He strapped himself into a canvas sling seat crowded against the bulkhead by the plane's cargo of thousand-pound electronic instruments and Hughes 500 helicopters. He was headed for Canada and the crash site, as part of an Atomic Energy Commission response team. "We knew that, intact, the reactor could give a lethal dose of radiation to anyone in its vicinity," Henry said.

Carl Henry was a physicist, a specialist in radiation detectors, who worked in the nuclear weapons program at Los Alamos National Laboratory. The soft-talking, baby-faced Henry, in his late thirties, was in good shape for an academic, but he was no mountaineer. He was used to more conventional surroundings: a metal desk, a small office with cream-colored walls. When he checked into the Igloo Hotel in the village of Baker Lake, the wind chill was 100° below zero. The sun came up at ten in the morning and set four hours later.

For several days, sniffer planes—aircraft fitted with radiation detectors—had made hundreds of sorties over the area of impact, but their readings were inconclusive. "Besides the snow, the only thing in the landscape were granite outcroppings," Henry said, "and they contained enough natural uranium and thorium to trigger the detectors." Carl Henry waited on the ground, ready to go in and handle any remnants when they were finally found.

As the planes crisscrossed the sky, a small group of adventurers, six young men, were camping in the treeless, snowbound plains along the Thelon River. They had seen the strange light of a fire the same night the satellite crashed. Two members of the party struck out on a dogsledding expedition and stumbled on a peculiar three-foot length of metal sticking out of the bottom of an ice crater. They radioed the information to Yellowknife. The Canadian Air Force dropped paratroopers in, posted guards at the site, and called in the science team.

The next day, Henry and a half-dozen other scientists left Baker Lake on a Chinook, a heavy-bodied, double-rotored helicopter. It was a 250-mile flight to Warden's Grove, where the scrap of satellite had burrowed into a frozen river. The piece of debris was a knot of metal with two prongs pointing sideways in the snow. They called it the "antlers." Swathed in parkas and mukluks, the team took measurements off the metal to gauge how much radiation the two dogsledders had received. There was not much danger from the radiation, but the cold was another matter. When the day was over, the helicopter would not start. A C-130 mother ship circled overhead. When it was clear the team was okay, it flew off, leaving the men to the quiet of a boundless night.

"We were stuck," Carl Henry said, "but we had flown in with a survival expert. We built snow walls and pitched tents. The Coleman stove used naphtha, the only fuel that didn't freeze at those temperatures.

"We had lots of canned food, but it was frozen solid. First we boiled the cans long enough to thaw them, then threw everything into one cooking pot: the chicken, the noodles, the peaches, everything we had."

They slept on the ice, under the soft, serpentine Northern Lights. Rescued the next day by a Twin Otter plane, they began a daily commute between the Igloo Hotel and the camp beside the Thelon River.

The next day they dug out the antlers and dumped everything they found into lead-lined garbage cans. "The antlers turned out to be the rods that controlled the reactor. We kept digging, thinking the reactor itself might be buried deeper in the ice, but we didn't find it," Henry said.

Warden's Grove would become a center for the ongoing search. Two weeks into it, the Canadian Air Force flew a C-130 three feet above the ice, opened the ramp, and deployed a parachute that dragged a bulldozer out onto the snow. The Caterpillar cleared a mile-long landing strip along the river, crashing through the delicate winter silence. Engineers wired up strobe lights and a navigation system to guide the air traffic.

Carl Henry had been plucked from the quiet, collegial halls of his laboratory and choppered in to these subarctic barrens because he was a member of the Nuclear Emergency Search Team, NEST. NEST was a secretive network of volunteer nuclear scientists and engineers that drew on expertise from Los Alamos and other national laboratories. They were trained to hunt down and defuse terrorist bombs, but their skills and special equipment were indispensable in any nuclear crisis.

Only a hundred pounds of uranium, smaller than a gallon of milk, not even a bomb but a nuclear engine, had marshaled the tumultuous assault on this frozen Arcadia. It was a paradox of nuclear numbers. In 1978, there were tens of thousands of fine-tuned nuclear weapons pointed at targets in the United States and the Soviet Union. It was a world bristling with nuclear-tipped arrows, electrified by anger. But the damage one errant lump of nuclear fuel could do was appalling, unacceptable, and had to be addressed with every tool at hand.

The scientists spread outward from Warden's Grove in ground parties, finding odd fragments here and there. Canadian planes logged more than thirteen hundred hours tracking radiation spikes. Search and cleanup operations continued far into the summer.

Henry said, "In the end, we figured out the reactor had come apart into pieces the size of pepper and scattered over ten thousand square miles."

He liked to talk about the ptarmigan, the fat, white snow grouse that flocked around Warden's Grove. They were the only creatures that had shared the wide, wintry plains with the pilots and physicists.

THE SEARCH AND CLEANUP IN CANADA, OPERATION MORNing Light, as it was called, was a minor incident among a multitude of East-West sword-rattlings, proxy wars, and near misses. But it foreshadowed the way all the nuclear dangers and all our nuclear priorities would turn bottom side up as the Cold War came to an end.

It was no longer a fusillade of fifty thousand warheads that threatened us; it was the the single unpredictable incident of negligence or the single smuggled bomb. It was no longer the fifty thousand nuclear targets scarring the face of the globe; it was the tons of plutonium sowed across the continents, a poison harvest.

The menace of nuclear war was not gone, it just had new front lines: Iraq, North Korea, South Africa, the Ukraine, Syria, Israel, the black market in plutonium, the underworld of political terrorism.

The men and women in national laboratories like Los Alamos, who knew the secret anatomy of the bomb, were the obvious place to look for solutions to the new nuclear problems. All the exuberance and sprawl of the Cold War technological buildup could be redirected toward the dangers lying in ambush in the new world. The weaponeers could become like Dr. Henry, nuclear firemen trained to race to contain any emergency. They could apply their brilliance to preventing the spread of nuclear weapons, and to reversing the spread of plutonium.

But adjustments to the changing scientific and military climate of the mid-1990s had to be made in a time of fiscal austerity, a time when raw scientific exploration was no longer an object of pride and fascination.

Dollars were short. Between 1981 and 1991, the United States spent $3 trillion on the military, and more than half of that money went toward facing down the Soviet Empire. We could not afford that kind of prodigality any longer. Federal budgets for basic research, pared back in 1992, were flat-lined through 1994. Applied science, the very practical kind that modernized semiconductors and designed robotics for manufacturing, was one of few areas to enjoy increases. Nuclear physics took a cut.

NASA, the great Cold War icon of American will and ingenuity, was adrift, the very image of an ossified, inept bureaucracy. Congress abandoned the Superconducting Supercollider, a Brobdingnagian particle physics experiment under construction in Texas, one that would have moved us closer and closer to knowing the origins of the universe and kept America in the forefront of physics.

There was less money for Los Alamos, the exemplar of Cold War science. Projects like magnetoencephalography, neutrino physics, and planetary exploration continued, but under the strains of uncertainty and budget pressure. Pure science was not a growth area at Los Alamos or anywhere. Without a mortal enemy as a goad, the public was losing interest in discovery.

As the stockpile shrank, funding for weapons research was dropping, down by 20 percent in 1992, with more cuts coming. The total number of employees engaged in weapons research, development, and testing at the lab was fewer than at any time since the dawn of the nuclear age. There were no new weapons being planned or built. At Fernald, Ohio, a site that had been processing uranium for nuclear warheads ceased operations. In 1994, all fifteen hundred production jobs at the plant were eliminated, while hazardous waste cleanup jobs doubled to over three thousand.

The assemblage of minds and matériel at Los Alamos was a national asset of great value. The question was how to use it well but not wastefully. It was necessary to stay on guard against mortal threats, but it was also necessary to be judicious in deciding what those threats truly were: a rogue nuclear state, a ravaged environment, a feeble economy? It was within the power of science to address all of these troubles.

THROUGHOUT WORLD WAR II, THE ROAD THAT LEFT THE Rio Grande valley at Otowi Bridge and rambled up to the top of Pajarito Plateau was narrow, unpaved, rutted, and unreliable in bad weather. The ascent was a natural hazing for anyone who would enter the secret society of the Manhattan Project.

In the summer of 1990, when the Soviet Union and the nuclear weapons complex was still intact, I made regular trips to the lab along the same route, now built to the standards of twentieth-century civilization. I used to see a stenciled slogan repeated on concrete abutments all along the way: DOE KILLS. I assumed it meant the Department of Energy, which administered the nuclear weapons program, was poisoning the environment with radioactive waste. I occasionally, too, saw protesters at the end of Omega Bridge, near the lab's main offices. They quietly paced, carrying ban-the-bomb signs.

By 1994, the roadside was free of antinuke propaganda. There was still writing on the walls, but it was the hasty, spray-painted scrawls of gang graffiti. Without the kill-or-be-killed mentality of the Cold War, the weaponeers had no obvious path to follow and even its critics stopped telling the lab what to do.

For Los Alamos, discovering the Cold War was over was like a slow awakening in a strange room. The question hung in the air: *Where are we?*

The careers of eight thousand workers, twenty-five hundred of them scientists and engineers, were captive to political turmoil and misdirection as the overseers of national security searched for a revised foreign policy and a downsized military. Everyone at the lab wondered if they would have a job in another year. Scientists started sending out résumés. There were retirements and a few layoffs. Workers shifted from division to division, or found small, out-of-the-way projects to keep them busy. Numbers in the core weapons program ended up half what they had been in the late eighties: nine hundred workers, down from eighteen hundred. Even the nonweapons sciences "outside the fence"— subatomic physics, planetary physics, biology, geology, astrophysics—had always been propelled by the urgency of Cold War competition. Their funding was hardly more reliable. Since 1990, the laboratory has shrunk to seven thousand employees.

In 1994, the lab's future was still ill-defined. An employee survey taken that year demonstrated that lab scientists were not at all happy with the idea that all they were doing was marking time, keeping the doors open. Management received embarrassingly low marks for "providing a clear sense of direction" and making "decisions consistent with the lab's values."

Any bureaucracy can be sluggish and self-defeating, but at Los Alamos there was more to the crisis of morale. The scientists who worked in the heart of the weapons program had always done their jobs because they believed that the right number and the right kind of nuclear weapons would save us from the savage global bloodletting of another world war. They would not be satisfied until they could replace that lost sense of purpose with something equally compelling. But the laboratory seemed just to be on standby. It had expended some effort on reorganizing management charts, and discussing new management styles, but the larger questions were neglected. A new motto was "Los Alamos, a Customer Focused, Unified Laboratory." What could that possibly mean?

No one was inventing new weapons anymore. Groups within the lab were looking around for the kind of problems that their solutions might fit. Weaponeers paused to assess whether nuclear deterrence was still the right thing to do. Management conjured up a new line of work they called technology transfer, an attempt to redirect at least some research toward commercial products like fuel cells, nonpolluting automobiles, and advanced semiconductors. Funding was shunted toward counterproliferation technology and stockpile stewardship, that is, monitoring weapons for safety and security.

Yet the lab's 1994 proposals for future funding included huge investments in research and development in nuclear weapons technology, a suggestion strangely out of sync with public perception, and even its own internal mood. The lab administration knew that these spending proposals were not going to look too palatable. A press officer attached a disclaimer to the projected budget, saying that a restatement of nuclear weapons policy from Washington could change the scope of their plans.

The real problem was that nobody was in charge. Los Alamos was buried in multiple, overlapping layers of responsibility. The lab director did not have anything like the power or discretion of a CEO. Lab management had to appease many masters. They were at the mercy of further instructions from their funding agency, the Department of Energy; from Congress; and from the Clinton administration, none of whom sent consistent messages. The nation had not decided exactly what to do with its nuclear weapons, except to get rid of a whole lot of them. Dismantling or redesigning the institutions that supported the weapons was an issue that was tabled every time it was raised. The mechanisms of change were clumsy, inadequate, and just generally hung up in traffic. Meanwhile, Los Alamos operated with an annual budget of around one billion dollars.

One weaponeer described the laboratory as a charging rhino. Its target was gone, but it had too much momentum to stop, too much poundage for quick turns. Meanwhile, lab workers clung to the back of the beast, not knowing where they would be at the end of the ride.

THERE WAS A BALD PATCH OF LAND ON FENTON HILL just off Route 4 in the Jemez Mountains, the site of Los Alamos's Hot Dry Rock experiment and David Duchane's laboratory. He was showing me around.

A wildfire blasted away the ponderosa in the early seventies. Afterward, the lab came in, cleared the deadwood, and laid down a field of gravel. The head scientist was a Sierra Clubber who saw to it that the surrounding land was reseeded. Wildflowers and young aspen had pushed past the fence and blurred the line of the perimeter, but the place still looked all business. Hot Dry Rock was a search for clean energy. Los Alamos had drilled two wells at Fenton Hill, steel tap roots driven more than two miles deep in the granite, probes thrust into the primeval heat of an old volcano.

Fenton Hill could have been a museum. Only a couple of technicians pottered around among the stainless steel piping. The machinery was silent. All you could hear was the wind in the hip-high grass that grew on the slopes nearby. The control room had a wall of twenty-year-old electronics, dark now. All its circuits and oscilloscopes had been replaced by a single desktop computer. The computer was turned off.

The experiment, though, was about to start up again. David Duchane had migrated out of weapons science into this small, isolated program, one of many such displacements that followed the fiscal pinch. Scientists who once collaborated on large, cohesive projects were scattering, adjusting to the changes by ducking into laboratory backwaters.

Dave's rescue came in the form of Hot Dry Rock— geothermal energy; it was mining the great store of heat from the furnace of the earth's interior. The experiment forced cool water into the 400-degree heart of the mountain, then drew it back up, hot enough now to drive turbines. This was electricity that would cost the environment almost nothing.

"We have tested it over and over, hundreds of times, and there are no emissions, no greenhouse gases, from the process," Dave said.

Back down the mountain in Los Alamos, Dave spent

his day tucked inside a cluster of portable buildings parked next to a fence with a radiation warning sign. I had learned that work space at the lab never gave anything away about the occupant's status in the hierarchy of science or in the labyrinth of management. The lab's indifference to ornament was due in part to the fact that it was a spender of public funds, accountable to visiting congressmen. Beyond that, it was the essence of experimental science to be expedient and single-minded. All I could tell from David Duchane's plain, seven-by-ten pigeonhole office was that he was tidy.

Dave was a trim fifty-two-year-old with Opie-of-*Mayberry* hair shooting out in sideways tangents. His hair was gray but I could still see the square-jawed, blue-eyed boy he used to be.

He grew up in Bay City, Michigan, a gritty brick-and-mortar town of shipyards and Chevrolet assembly lines. His mother was a waitress and his father drove trucks and worked in a gas station. He was the only one among his friends to go to college right out of high school.

College was not really what Dave had in mind for himself either. He bounced in and out of school until he earned his Ph.D. at the age of thirty-six and came to Los Alamos.

The lab offered him a well-paid job and the wide-open splendor of the West. He started in the materials science division, where custom-made components for weapons tests and laser fusion experiments were invented and fabricated in a hands-on laboratory.

Dave was a born liberal. He had joined the McGovern campaign in 1972 because he was against the war in Vietnam. He knocked on doors and tried to talk what he called "the owning class" into voting Democratic. He was never uncomfortable working on bombs. He could see the vertiginous logic of deterrence. "Look," he said, "everyone here knows the weapons really work. So we are the last people who would ever want to see them used."

Dave had sidestepped the consequences of the end of the Cold War by finding his own place to do good works.

"Hot Dry Rock could be the second biggest thing the lab has ever done. It would be a new source of energy, one important solution for the future," he said.

Enormous amounts of energy waited beneath the surface of the earth. The U.S. Geological Survey estimated that all the natural heat stored in the first four miles of rock under the forty-eight states was equivalent to 3,000 trillion barrels of oil.

Dave imagined geothermal plants dotting the country, feeding clean energy into the power grid. He pictured skyscrapers with heat and light from their own geothermal pumps in the basement. Other scientists at the lab would say, *Hot Rock, are we still doing that?* The project had been around for twenty years, at one time operating with a $20 million annual budget. In 1994, funding was down to $1.25 million. For all its potential, the project gave off the scent of defeat.

"I've got some support, but my project is not seen as the future of the lab," Dave said. Nevertheless, he thought of himself as a lucky man. If the administration was not entirely helpful, at least it was benevolent. He was looking forward to a small increase in funding.

It was going to take something like Dave's zeal to get the job done. Working inside a big bureaucracy like the Los Alamos lab tended to dilute one's sense of power over outcomes. "You spend as much time justifying what you are doing as you spend actually doing it," Dave said. "It's different from working for a business. Here there is always a paper trail, a rule book, a lot of Thou Shalt Nots. Here's an example. When I was in the research lab at Kimberly-Clark, the purchasing agents' job was to get you what you needed to do your work; here, the job is to make sure purchases meet all federal requirements.

"A group leader cannot promote a technician until he fills out an eight hundred-item questionnaire. How can you do that sensibly? At Kimberly-Clark, the boss actually appeared to be in charge. You were more likely to do what he asked you to do because you knew he was the guy who was absolutely responsible for what happened to you."

I had heard variations on Dave's criticisms from every single staff member I had ever met. But all the red-tape aggravation was tolerable as long as the staff felt that what they were doing made sense. Dave knew he had found an island of substance and meaning within Los Alamos, while his friends were still at sea. He understood the uneasiness all around him.

Dave said this: "No one expects nuclear weapons to make a comeback. No one expects another nuclear test. We need a legitimate new basis for broad public support. We can't simply say we will do great science. There are hundreds of federal labs and thousands of universities doing good science. We need our own niche. The logical niche is for the lab to continue with its specialty, nuclear phenomena. We could also do nuclear cleanup, and just shrink back to a core of weapons work, one without innovative science. Nuclear weapons don't need innovation anymore."

Los Alamos scientists know very well that outsiders are suspicious of everything they do. "If we work on the cleanup of nuclear waste, people will think we are doing it as an excuse to make more bombs or more nuclear power plants, only cleaner ones," Dave said. "The trouble is, anything we try will look, in the short term, like we are just doing it for the money, for our own survival and not for the good of the country."

FOR NEARLY THREE YEARS, JAS MERCER-SMITH AND I had been meeting regularly in the lab library, but some time had passed since we last spoke. In the fall of 1993, he had been at low ebb, faced with the end of nuclear testing and 30 percent staff cuts in his division. Two-thirds of the staff had signed up to put time in on global climate modeling.

Now, a few months into 1994, he was invigorated by new prospects. Within the design division, certain tasks had become the obvious ones to pursue. "The more people there are with weapons, the more likely nuclear war becomes," Jas said. "We designers had to ask ourselves, how could we contribute to nonproliferation?"

The prospect of an outlaw nation with a nuclear arsenal appalled him. "Genghis Khan started with nothing more than a few *horses* and he conquered all of Asia. He committed genocide to make the world safe for goatherders. Holocaust and self-delusion—that's human nature. And here I have been designing nuclear weapons. Well, the world has changed, so deterrence changes, but that just means you have to adapt and keep going."

Los Alamos was devoting more money and manpower to nonproliferation technology like satellite surveillance.

"We can help stop or at least slow nuclear proliferation," Jas said. "For instance, one woman in my group has posed this question: If I were a rogue weapons designer, how would I bypass export controls to bring the parts and materials I need into, say, Iran? What kind of trail would those tricks leave? An expert would behave differently from a beginner."

More than one shipment of contraband plutonium had been seized in Germany that year. Physicists at a lab near Munich, using the most rigorous analysis, had been able to establish that at least one batch came from a research reactor in Russia. This kind of detective work was also a part of counterproliferation, what Jas called nuclear forensics.

"We need technology that can tell us exactly who blew up a bomb. When it happens, we need to be able to say who did it—in fifteen minutes. The ability to do that would be a deterrent to renegades."

The lab continued to be charged with the statutory requirement to tend to the nuclear stockpile and maintain the capability to build more if necessary, the idea being you can't turn your back on a dragon even when it is sleeping. Jas was expected to keep a small cadre of thermonuclear designers trained at a level of peak fitness in case the need to design new weapons arose again. No one was quite sure how that could be done in the absence of nuclear testing. Jas said, "We will try to devise some piecemeal nonnuclear tests to look at small parts of the process."

The design division was confronting another problem. Like everything else in the world, nuclear weapons age. They crack, chip, corrode, expand, and contract. "There is some nasty stuff inside a weapon," Jas said. "Hydrogen embrittles metal. Neutrons disrupt all kinds of materials. And each type of weapon is going to get old in its own specific way."

With no nuclear weapons in production and the

factories that made bomb parts shutting down, keeping a watch over the stockpile had a whole new meaning.

"These weapons were designed to last about twenty years," Jas said. "We have no idea how they will behave after fifty years." The nuclear stockpile was a deterrent only as long as you knew it was in working order. "The fewer bombs we have," he said, "the more important each one becomes.

"Analyzing the aging is a stochastic problem. It is technically much harder than designing the bombs in the first place. To do the job well with the computers we have now is like trying to cut down a sequoia with a nail file. But as long as we have nuclear weapons, it is my job to guarantee their safety and reliability," he said.

"Somebody asked me the other day, 'Isn't it hubris to think they are still deterring anything?' My answer was, 'I may be a fool for thinking that deterrence will keep working, but you better hope I'm not.'"

SUNBATHING IS ONE OF THE FEW UNHEALTHY PERSONAL practices at the lab. Almost nobody smokes and every bulletin board invites employees to exercise at The Wellness Center. At the front door of the cafeteria is a machine to measure pulse and blood pressure; inside, the menu offers lots of low-fat options. Larry Madsen was sitting on a patio, squinting in the full sun of the summer of 1994 when I asked him, "What are you doing now?"

"I'm not sure," he said, "I don't know what role I have in the weapons program anymore, or what the program is going to be."

He, more than any of the Los Alamos scientists I have gotten to know over the years, sees his job's sudden lack of coherence as a crisis of conscience. The lab's scramble for self-justification is no substitute for genuine service in the national interest. "We've never had pork-barrel status before," he said. "And it is hard to get people here to look beyond the current tactics of survival.

"What does the public want? They want peace, and they want a good defense. They want a national security insurance policy, but they don't want to pay for it with too much environmental mess. You can write all this down on the wall and then build yourself a program. Say 'I'm trying to eliminate that,' or 'I'm trying to look at this problem, that's what I'm doing with your tax money.' I could ask, why aren't we actively trying to figure out how to develop a class of weapons that eliminates the need for testing, or plutonium, or tritium. Why wouldn't that be something the public wants? Why can't we take on the problems that are so far out nobody else will work on them? Like becoming the research and development arm of law enforcement; the Environmental Protection Agency doesn't have an R & D arm.

"We are using government money. Maybe what we should do are things that nobody else wants to do. We could work on technology for law and order, or transportation, or looking out for the environment.

"If the argument is that we need to maintain our nuclear competency, I'm not sure that watching the stockpile and worrying is enough.

"The lab just says as long as nuclear weapons are around, we've got to be around. I keep arguing that we are asking for more money but we are not solving anybody's problem.

"Sometimes I think I am hollering in a tunnel. There is no forum for discussing change. Nobody owns the big question," Larry said. "What are the national labs for? It's everybody's job to answer that, not just senior management, not some bureaucrat making decisions from on high."

It was an echo of something Del Bergen had said when the Cold War had just ended: "The lab has no right to rely on its history to maintain itself now. It is time for taxpayers to figure out how science should serve national security. And it's time to expand that notion to include economic security and energy security along with military."

John Hopkins, who had finished his year in San Diego at the Institute for Global Conflict and Cooperation, retired from the lab in 1994. He, too, was unsentimental about the scientific establishment. "It is big and too inefficient. It should be downsized," he said, "just like the Army and Navy. There is little the lab can do that can't be done as well somewhere else."

Without an overarching goal, Los Alamos has nothing to offer the body politic that other institutions cannot also offer. "Government is incredibly inefficient at operating big science," he said. "If we have a unique capability, that is the only reason to keep a science going here. We should do the Human Genome Project, for example, because of our supercomputing power. But overall, I can imagine the lab at half the current workforce."

As for counterproliferation technology, Hopkins said, "Nonproliferation is a political problem, not a technical one. The direction to go is toward an international agreement that makes nuclear weapons illegal. That would give us the status to enforce export controls, to monitor nuclear weapons material, and to consider military action against a proliferator. If we outlaw weapons, there would be no need for weapons labs."

IGCC's founder, Herbert York, a wise man who had grown up and grown old in the nuclear weapons culture believed that the Cold War did not end because of our nuclear weapons but because of the West's political and economic architecture: democracy, stable currencies, free trade. "We didn't win the war," he said. "We won the peace."

THERE IS MORE LIGHTNING IN NEW MEXICO THAN ANYwhere else in the forty-eight contiguous states and a wall of storm clouds to the east was rushing toward us. Since I was standing six feet away from a hundred pounds of high explosives, I asked if a lightning strike might be a problem and Steve Younger said, "You'd have to worry about that, yes."

Steve and I were in Ancho Canyon, a narrow crack between two mesas, on a test range where Los Alamos scientists blew things up. We were watching a team of technicians from Arzamas-16, Russia's nuclear weapons laboratory, rig a metal capsule containing deuterium to the end of a long tube that held a copper pipe filled with chemical explosives. The pipe was surrounded by coils for carrying millions of volts of electricity.

It was an experiment designed to use this intense electromagnetic energy to try to produce controlled fusion with the kind of pressures and temperatures at the heart of a star—ten million degrees of heat for a few microseconds. The apparatus rested on a cradle of two-by-fours set up under a flimsy metal roof. Plastic tarps hung up with bright yellow tape formed the sides of the temporary housing—temporary because the whole setup was going to be blasted into oblivion when the experiment went off.

American and Russian physicists stood around under the darkening sky watching a workman hack off an awkward edge of wood. At this point, the guy who knew how to screw the parts together was in charge and the Ph.D.s were just onlookers.

Lines of sandbags wound around and up the side of the mesa. There were thick metal plates dug in and canted to shield some camera equipment. Cables run along the ground tethered the experiment to all the electronic monitors set up in the bunker nearby.

Steve introduced me to Vladimir Konstantinovich Chernyshev, an old-time weaponeer who had developed high-explosive components for Russian thermonuclear bombs. Chernyshev was a tall, thin-lipped man with a distracting number of ruts and nodules on his face. He spoke rather formally, as Russians tend to do with journalists. "It is important to keep weapons scientists busy with intriguing, demanding science," he said. "This experiment is ideal."

At that moment, we noticed a brown tarantula making its deliberate way off the hill and onto a sandbag. Chernyshev's attention turned to its capture. He brought it back a few minutes later in a paper cup. "I'll take it to my grandson," he said.

The thunderheads had cleared off by one o'clock, about the time the experiment was ready. Everyone, around fifty of us, crowded into the concrete bunker bermed into the mesa below the firing point. After a roll call and safety briefing, Steve parked me next to a television screen to watch the explosion. "Keep your distance from any metal," he said, "and cover your ears."

We all waited, and I watched the kinetic tension among the men who stood at their posts where the first results would appear as jagged lines on a dozen different monitors. "This is a new configuration," Steve said, "and we just don't know whether our high explosives are going to work with the Russian machine."

IN 1992, IN AN ULTIMATE EXPRESSION OF PEACEFUL intentions, the Russian and American weapons laboratories had opened their doors to one another. Visits between the former enemies went beyond symbolism: the idea was to fashion a direct collaboration in nonweapons research and bring some badly needed funds to Russian scientists.

Steve Younger was party to one of the earliest exchanges between the labs. He was in awe of some of the Russian achievements. "I said to myself, I don't care if they're from Russia or Argentina or Mars, we've got to work with these people."

The collaboration was slow going. "Some people in our government didn't want it. They said the money we put into it would end up being used against us," Steve told me. "One less than sober Russian I ran into said, 'I spit on your generosity.'"

Steve decided to fight his way through the resistance.

Younger was an immigrant from Los Alamos's sister laboratory, Livermore. He took a job there after calling the 800 number from an ad in the Sunday paper: "Physicists Wanted for Exciting and Creative Project." He designed nuclear weapons for a while, then moved on to x-ray lasers, a Star Wars project. After losing faith in that program, he came to Los Alamos, where he worked his way up to Deputy Director for Nuclear Weapons Technology.

Steve had unremarkable features and fair, thinning hair. He was middle everything: height, weight, age. He had presence, though. Like most weapons designers, he has an abundance of self-confidence, but enough wit to offset any traces of arrogance.

Steve eventually prodded the two reluctant sides into

signing an agreement to cooperate on this series of pulsed power experiments that meant American dollars for Russian weaponeers. "The Arzamas-16 community was growing their own food to sustain themselves through winter. They had not been paid for three months. We said, 'Look, we've got a lot of people here who know a lot about nuclear weapons. They are not going to start making bicycles. There are countries willing to pay them a great deal of money just to come sit on the faculty of a national university.'"

There were reports of fifty Russian nuclear specialists living in Iraq and fourteen in Iran.

"It was a national security issue for us to see to it that Arzamas got some support," Steve said. "Besides, they were five years ahead of us in pulsed power research. There was something to gain."

AT 1:07 P.M. ON OCTOBER 7, 1994, THE PULSED POWER experiment went off in Ancho Canyon, exactly according to plan. I saw a dazzling orange billow of flame on the video screen. The floor jumped and I heard a low pop. An acid-burned odor drifted into the bunker moments later.

Almost immediately, the doors were opened, and technicians swarmed out to pick through the ruins. Inside the bunker, a dozen scientists stood in a huddle around a computer, talking excitedly over and under each other. They could hardly believe it, but it looked like they had achieved a flash of nuclear fusion.

I climbed back up to the site, and stared at the empty place where the experiment and the surrounding structures had been only minutes before. Twisted scrap and scorched shrapnel had sprayed across the hillside. There was a thrill to the absoluteness of the destruction.

"Some explosions light the tops of the trees like candles," Steve said.

I had just witnessed something I would have never expected to see: a Russian fusion device on American soil, detonated, not for killing, but for the fabulous mischief of making star fire.

ABOUT THE AUTHOR

JANET BAILEY writes about science for numerous publications, including the *Los Angeles Times*. She is the author of *Chicago Houses* and *Keeping Food Fresh*. She and her husband own Barbara's Bookstores, a small chain located in the Chicago and Boston areas. She lives with her family on a postage-stamp ranch in Santa Fe, New Mexico, close enough to keep her eye on neighboring Los Alamos.